SCIENCE AND TECHNOLOGY FOR ARMY HOMELAND SECURITY

REPORT 1

AF754876

Committee on Army Science and Technology for Homeland Defense
Board on Army Science and Technology
Division on Engineering and Physical Sciences

NATIONAL RESEARCH COUNCIL
OF THE NATIONAL ACADEMIES

THE NATIONAL ACADEMIES PRESS
Washington, D.C.
www.nap.edu

THE NATIONAL ACADEMIES PRESS 500 Fifth Street, N.W. Washington, DC 20001

NOTICE: The project that is the subject of this report was approved by the Governing Board of the National Research Council, whose members are drawn from the councils of the National Academy of Sciences, the National Academy of Engineering, and the Institute of Medicine. The members of the committee responsible for the report were chosen for their special competences and with regard for appropriate balance.

This study was supported by Contract/Grant No. DAAD19-02-C-0049, TO 2, between the National Academy of Sciences and the Department of the Army. Any opinions, findings, conclusions, or recommendations expressed in this publication are those of the author(s) and do not necessarily reflect the views of the organization that provided support for the project.

International Standard Book Number 0-309-08701-5

Cover: The Pentagon burning after being struck by a commercial airliner, September 11, 2001. Courtesy of Reza Marvashti, The Free Lance-Star, Fredericksburg, Virginia.

Additional copies of this report are available from the National Academies Press, 500 Fifth Street, N.W., Lockbox 285, Washington, DC 20055; (800) 624-6242 or (202) 334-3313 (in the Washington metropolitan area); Internet, http://www.nap.edu

Copyright 2003 by the National Academy of Sciences. All rights reserved.

Printed in the United States of America

THE NATIONAL ACADEMIES

Advisers to the Nation on Science, Engineering, and Medicine

The **National Academy of Sciences** is a private, nonprofit, self-perpetuating society of distinguished scholars engaged in scientific and engineering research, dedicated to the furtherance of science and technology and to their use for the general welfare. Upon the authority of the charter granted to it by the Congress in 1863, the Academy has a mandate that requires it to advise the federal government on scientific and technical matters. Dr. Bruce M. Alberts is president of the National Academy of Sciences.

The **National Academy of Engineering** was established in 1964, under the charter of the National Academy of Sciences, as a parallel organization of outstanding engineers. It is autonomous in its administration and in the selection of its members, sharing with the National Academy of Sciences the responsibility for advising the federal government. The National Academy of Engineering also sponsors engineering programs aimed at meeting national needs, encourages education and research, and recognizes the superior achievements of engineers. Dr. Wm. A. Wulf is president of the National Academy of Engineering.

The **Institute of Medicine** was established in 1970 by the National Academy of Sciences to secure the services of eminent members of appropriate professions in the examination of policy matters pertaining to the health of the public. The Institute acts under the responsibility given to the National Academy of Sciences by its congressional charter to be an adviser to the federal government and, upon its own initiative, to identify issues of medical care, research, and education. Dr. Harvey V. Fineberg is president of the Institute of Medicine.

The **National Research Council** was organized by the National Academy of Sciences in 1916 to associate the broad community of science and technology with the Academy's purposes of furthering knowledge and advising the federal government. Functioning in accordance with general policies determined by the Academy, the Council has become the principal operating agency of both the National Academy of Sciences and the National Academy of Engineering in providing services to the government, the public, and the scientific and engineering communities. The Council is administered jointly by both Academies and the Institute of Medicine. Dr. Bruce M. Alberts and Dr. Wm. A. Wulf are chair and vice chair, respectively, of the National Research Council.

www.national-academies.org

COMMITTEE ON ARMY SCIENCE AND TECHNOLOGY FOR HOMELAND DEFENSE

JOHN W. LYONS, NAE, *Chair,* U.S. Army Research Laboratory (retired), Mount Airy, Maryland
GEORGE BUGLIARELLO, NAE, Polytechnic University, Brooklyn, New York
TIMOTHY COFFEY, University of Maryland, College Park, with joint appointment at National Defense University, Washington, D.C.
STEPHEN W. DREW, NAE, Princeton University, Princeton, New Jersey
MITRA DUTTA, University of Illinois, Chicago
FREDERICK L. FROSTIC, Booz Allen Hamilton, McLean, Virginia
C. WILLIAM GEAR, NAE, NEC Research Institute, Princeton, New Jersey
ARTHUR H. HEUER, NAE, Case Western Reserve University, Cleveland, Ohio
HOWARD S. LEVINE, Weidlinger Associates, Inc., Los Altos, California
JOSEPH P. MACKIN, E-OIR Measurements, Inc., Spotsylvania, Virginia
JACK N. MERRITT, U.S. Army (retired) and Association of the U.S. Army (retired), Arlington, Virginia
THOMAS E. MITCHELL, Gray Hawk Systems, Inc., Alexandria, Virginia
K. DAVID NOKES, Sandia National Laboratories, Albuquerque, New Mexico
DENNIS J. REIMER, U.S. Army (retired) and Memorial Institute for the Prevention of Terrorism, Oklahoma City
EUGENE SEVIN, NAE, Consultant, Lyndhurst, Ohio
ANNETTE L. SOBEL, Sandia National Laboratories, Albuquerque, New Mexico
MICHAEL F. SPIGELMIRE, U.S. Army (retired), Consultant, Destin, Florida

Liaison, Board on Army Science and Technology

DONALD R. KEITH, U.S. Army (retired) and Cypress International (retired), Alexandria, Virginia

National Research Council Staff

MARGARET N. NOVACK, Study Director
JAMES C. MYSKA, Research Associate
TOMEKA N. GILBERT, Senior Project Assistant

BOARD ON ARMY SCIENCE AND TECHNOLOGY

JOHN E. MILLER, *Chair*, Oracle Corporation, Reston, Virginia
GEORGE T. SINGLEY III, *Vice Chair*, Hicks and Associates, Inc., McLean, Virginia
ROBERT L. CATTOI, Rockwell International (retired), Dallas, Texas
RICHARD A. CONWAY, NAE, Union Carbide Corporation (retired), Charleston, West Virginia
GILBERT F. DECKER, Walt Disney Imagineering (retired), Glendale, California
ROBERT R. EVERETT, NAE, MITRE Corporation (retired), New Seabury, Massachusetts
PATRICK F. FLYNN, NAE, Cummins Engine Company, Inc. (retired), Columbus, Indiana
HENRY J. HATCH, NAE, Army Chief of Engineers (retired), Oakton, Virginia
EDWARD J. HAUG, University of Iowa, Iowa City
GERALD J. IAFRATE, North Carolina State University, Raleigh
MIRIAM E. JOHN, California Laboratory, Sandia National Laboratories, Livermore
DONALD R. KEITH, U.S. Army (retired), Cypress International (retired), Alexandria, Virginia
CLARENCE W. KITCHENS, IIT Research Institute, Alexandria, Virginia
SHIRLEY A. LIEBMAN, CECON Group (retired), Holtwood, Pennsylvania
KATHRYN V. LOGAN, Georgia Institute of Technology (professor emerita), Roswell
STEPHEN C. LUBARD, S-L Technology, Woodland Hills, California
JOHN W. LYONS, NAE, U.S. Army Research Laboratory (retired), Mount Airy, Maryland
JOHN H. MOXLEY, IOM, Korn/Ferry International, Los Angeles, California
STEWART D. PERSONICK, Drexel University, Philadelphia, Pennsylvania (until December 31, 2002)
MILLARD F. ROSE, Radiance Technologies, Huntsville, Alabama
JOSEPH J. VERVIER, ENSCO, Inc., Melbourne, Florida

Staff

BRUCE A. BRAUN, Director
MICHAEL A. CLARKE, Associate Director
WILLIAM E. CAMPBELL, Administrative Officer
CHRIS JONES, Financial Associate
DANIEL E.J. TALMAGE, JR., Research Associate
DEANNA P. SPARGER, Senior Project Assistant

Preface

This study is being conducted by the Committee on Army Science and Technology for Homeland Defense of the Board on Army Science and Technology, in the Division on Engineering and Physical Sciences of the National Academies. Sponsored by the Deputy Assistant Secretary of the Army for Research and Technology, the committee will produce a series of reports encompassing possible science and technology in support of the Army's role in homeland security (HLS). The statement of task for this first report is as follows:

> The National Research Council will:
>
> Review relevant literature and activities, such as the National Academies' emerging Science and Technology Program plan and Research Strategy for Combating Terrorism and their work with the interagency Technical Support Working Group (TSWG), reports from the Gilmore Commission and Hart-Rudman Commission, the DoD Counter-Terrorism Technology Task Force (DCT3F) plan, DOD Information Assurance policies and existing military operation and contingency plans to develop an Army context for the enhanced campaign against terrorism.
>
> Determine areas of emphasis for Army S&T in support of counterterrorism (CT) and anti-terrorism (AT). Operational areas the NRC should examine include indications and warning, denial and survivability, recovery and consequence management, and attribution and retaliation.
>
> In the first year, produce a report within nine months from contract award containing findings and recommendations that provide insights for high-payoff technologies.

BACKGROUND OF THE STUDY

The terrorist attacks of September 11, 2001, have forced the nation to consider how to prepare for the defense of the homeland. Terrorism is no longer an item on the evening news, taking place in some distant locale. Terrorism has become a domestic issue. As part of this recognition, the Army requested that the Board on Army Science and Technology (BAST) create a committee to meet over a 3-year period to consider how science and technology might better enable the Army to accomplish its mission in the homeland. It is anticipated that the committee will produce several reports during this period.

COMMITTEE PROCESS

This first report is a broad survey of relevant technologies, written in a relatively short period of time. Because of the scope of the review, the lack of a well-defined operational framework,[1] and the time-sensitive nature of the Army's interest, the committee has determined not to study specific products but rather to consider areas of technologies one level above individual products, processes, or services. In any case it should be noted that it is not the intent of this study to recommend budget actions; the technology assessments are intended to assist the Army in formulating its future technology plans.

The committee began its work by reviewing the literature listed below but found that very little has been said about the Army's role in HLS and the technology needs in support thereof.

- The National Strategy for Homeland Security,
- The Federal Response Plan,
- The National Academies' report *Making the Nation Safer: The Role of Science and Technology in Countering Terrorism,*
- The interagency Technical Support Working Group (TSWG) outputs,
- Reports from the Gilmore Commission and the Hart-Rudman Commission,
- The Department of Defense (DoD) Counter-Terrorism Technology Task Force (DCT3F) plan,
- DoD information assurance policies, and
- Existing military operation and contingency plans.

There are other reports, such as the annual report of the Department of Energy's Chemical/Biological National Security Program (CBNP), that the committee did not review for lack of time but that might provide additional information to the reader.

[1]Operational framework refers to a plan that the Army would use to conduct whatever operation may be necessary in response to a terrorist attack.

In addition to the literature search, the committee requested a series of briefings from the Army to better understand the Army's view of the homeland mission. It also heard from representatives of the National Guard Bureau to understand the role of the Army National Guard. A thorough legal briefing on the limitations of the Posse Comitatus Act facilitated this understanding. Lastly, the committee heard from scientists with expertise in a wide range of technologies in an effort to preview emerging types of equipment.

Even as this report was being prepared, doctrine and policy were being developed. The Department of Homeland Security and the Department of Defense's Northern Command, which are to have the major responsibilities and authorities for homeland security at the national level, are still in the early stages of formation and organization. The actual role that will be played by the Army in homeland security must certainly depend in large measure on the operational assignments Army units will be given in the framework of, or in support of, these overarching organizations. This remains in a state of flux. While, as is indicated in the report, it is anticipated that much of the doctrine will be drawn from existing protocols, the lack of specific doctrine made the study of specific equipment requirements difficult. Therefore the committee assumes certain functional requirements, which are described in Chapter 1.

REPORT ORGANIZATION

The DOD's Defense Counter-Terrorism Technology Task Force (DCT3F), in calling for and reviewing technical proposals in the wake of September 11, used the following taxonomy:

- Indications and warning,
- Denial and survivability,
- Recovery and consequence management, and
- Attribution and retaliation.

The study sponsor chose to make this taxonomy the basis for the committee's tasking document,[2] so the report is organized around these operational areas.

[2]In other documents, the Pentagon has used a different taxonomy but to the same end. For example, the Joint Warfighting Science and Technology Plan uses the following groupings of operational capabilities and subcapabilities:

Prevention	*Protection*	*Response*
Denial	Infrastructure	Attribution
Indications and warnings	Personnel	Consequence management
Deterrence	Facilities	Crisis management
Preemptive strike	Retaliation	

These four areas describe events in a time continuum beginning when intelligence indicates an event may take place and ending when blame can be attributed and appropriate retaliation executed. In Chapters 2 through 5 the committee has divided the four operational areas first into functional capabilities and then into technologies. Because the same technologies may be necessary in more than one of the operational areas, conclusions and recommendations concerning these technologies may appear in more than one chapter. Chapter 6 captures the overarching observations of the committee and Chapter 7 lists the findings, conclusions, and recommendations.

COMMITTEE COMPOSITION

The membership of this committee was intended to contain a broad representation of scientific and technological skill sets that have application to the Army's role in homeland security. These skill sets range from information technologies such as communications, computer sciences, and sensor technologies to materials and civil engineering, with special emphasis on structural hardening and resistance to nuclear and conventional explosive forces. Biosecurity expertise was considered important, as was a thorough understanding of the Army's capabilities. A security clearance was considered essential, as many of the topics that would be of interest to the committee are classified.

The committee worked very hard at its task and is grateful to all those who contributed to the report. Although the report limits itself to a fairly high-indenture level of exploration, the committee is satisfied that it will provide significant assistance to the Army as it moves on to future missions.

John W. Lyons, *Chair*
Committee on Army Science and
Technology for Homeland Defense

Acknowledgment of Reviewers

This report has been reviewed in draft form by individuals chosen for their diverse perspectives and technical expertise, in accordance with procedures approved by the NRC's Report Review Committee. The purpose of this independent review is to provide candid and critical comments that will assist the institution in making its published report as sound as possible and to ensure that the report meets institutional standards for objectivity, evidence, and responsiveness to the study charge. The review comments and draft manuscript remain confidential to protect the integrity of the deliberative process. We wish to thank the following individuals for their review of this report:

Thomas N. Burnette, Jr., LTG U.S. Army (retired),
Ashton B. Carter, Harvard University,
Anthony Dirienzo, Colsa Corporation,
Ronald O. Harrison, MG, Army National Guard (retired),
J. Jerome Holton, Defense Group Inc.,
Michael R. Ladisch, NAE, Purdue University,
Lewis E. Link, LTG, U.S. Army Corps of Engineers (retired),
John E. Miller, Oracle Corporation,
M. Allan Northrop, Microfluidic Systems, Inc.,
George W. Parshall, NAS, E.I. du Pont de Nemours & Company,
Harvey W. Schadler, NAE, GE Corporate Research and Development, and
Andrew Sessler, NAS, Lawrence Berkeley National Laboratory Center.

Although the reviewers listed above have provided many constructive comments and suggestions, they were not asked to endorse the conclusions or recommendations nor did they see the final draft of the report before its release. The review of this report was overseen by Alexander H. Flax, NAE. Appointed by the NRC's Report Review Committee, he was responsible for making certain that an independent examination of this report was carried out in accordance with institutional procedures and that all review comments were carefully considered. Responsibility for the final content of this report rests entirely with the authoring committee and the institution.

Contents

Tables, Figures, and Boxes

TABLES

FIGURES

BOXES

Acronyms

2-D	two-dimensional
3-D	three-dimensional
A and R	attribution and retaliation
AMC	Army Materiel Command
ARNG	Army National Guard
ATD	Advanced Technology Demonstration
BCT	brigade combat team
C&C	computer and communications
C2	command and control
C4ISR	command, control, communications, computers, intelligence, surveillance, and reconnaissance
CBR	chemical, biological, and radiological
CBRN	chemical, biological, radiological, and nuclear
CBRNE	chemical, biological, radiological, nuclear, and high explosive
CM	consequence management
CM and R	consequence management and recovery
CST	civil support team
D and S	denial and survivability
D2PC	Dispersion and Diffusion Puff Calculator
DARPA	Defense Advanced Research Projects Agency

DASA (R&T)	Deputy Assistant Secretary of the Army for Research and Technology
DHS	Department of Homeland Security
DoD	Department of Defense
DOE	Department of Energy
DTRA	Defense Threat Reduction Agency
EMT	emergency medical team
EPA	Environmental Protection Agency
ESF	emergency support function
FBI	Federal Bureau of Investigation
FCO	federal coordinating officer
FEMA	Federal Emergency Management Agency
FIOP	Family of Integrated Operational Pictures
FRERP	Federal Radiological Emergency Response Plan
GPS	Global Positioning System
HHS	Department of Health and Human Services
HLS	homeland security
HVAC	heating, ventilation, and air conditioning
I and W	indications and warning
ID	identification
IEW	intelligence and early warning
IR	infrared
JIC	Joint Information Center
JOC	Joint Operations Center
LFA	lead federal agency
LVB	large vehicle bomb
LWIR	long-range infrared
NCP	National Oil and Hazardous Substance Pollution Control Plan
NORTHCOM	Northern Command
OPSEC	operational security
OSC	on-site coordinator
PCA	Posse Comitatus Act
PDD	Presidential Decision Directive
ppb	parts per billion

ppm	parts per million
ppt	parts per trillion
R and CM	recovery and consequence management
R&D	research and development
ROC	regional operation center
S&T	science and technology
SBCCOM	U.S. Army Soldier and Biological Chemical Command
SCADA	supervisory control and data acquisition
SNR	signal-to-noise ratio
TRL	technology readiness level
TSWG	Technical Support Working Group
UAV	unmanned air vehicle
UGS	unattended ground sensors
USACE	U.S. Army Corps of Engineers
USAR	U.S. Army Reserve
UV	ultraviolet
VLSTRACK	vapor, liquid, and solid tracking
WMD	weapon(s) of mass destruction

Executive Summary

The U.S. Army is facing a challenge. At the same time that it launches a transformation toward the futuristic Objective Force, the centuries-old requirement to support civil authorities has been brought to the fore by the terrorist attacks of September 11, 2001. As the Army prepares for its still-evolving role in homeland security (HLS), the National Research Council was requested to establish a study committee under the Board on Army Science and Technology to advise the Army on how science and technology (S&T) could assist in the conduct of HLS. This is the first report from the committee.

This executive summary follows the same organization as the report. The section on background abstracts Chapter 1, where the context for the HLS mission is developed. The remainder of the summary addresses the technologies required over the four operational areas identified by the sponsor:

- Indications and warning,
- Denial and survivability,
- Recovery and consequence management, and
- Attribution and retaliation.

The technologies are displayed in tabular format in Chapters 2-5. Such a format provides the best way to understand the technologies the committee believes are important. A summary table depicting high-payoff technologies is provided at the end of this executive summary and in Chapter 6.

The main observations of this report are as follows:

- The S&T required by the Army for HLS need not be unique. The S&T work already being done for the Objective Force could provide much of the technology needed for HLS. In fact, if approached properly, the HLS effort not only can advance the S&T needed for the Objective Force, but also can assist in developing tactics, techniques, and procedures.
- The Army National Guard is critical to the success of the Army's efforts in HLS.

BACKGROUND

Homeland Security Requirements

While the operational framework[1] for combating terrorism on U.S. soil is still emerging, it is clear that this framework will be national in scope and based on cooperation. Although all disasters—either manmade or natural—are local, any disaster of great magnitude will require close cooperation among federal, state, and local governments. In case of a terrorist attack, the wide-ranging capabilities of our armed forces will most certainly be called on. The Army will have to cooperate with civilian emergency responders in order to save lives and mitigate damage. The Army's notional plan for HLS separates high-intensity homeland defense scenarios from lower-intensity civil support scenarios.

The military is not the only community seeking to learn from the events of September 11. The committee became aware of ongoing efforts in the civil sector to develop equipment for civilian emergency responders. This commercially developed equipment might have great applicability for the Army, but there does not appear to be a mechanism for integrating the research being done in the civilian community with that being done in the military community.[2]

Recommendation. The Army should encourage better coordination of the disparate homeland security science and technology efforts.

Recommendation. The Army should facilitate technology transfer in order to allow the private sector and other government agencies to exploit the homeland security technologies it develops.

---

[1]Operational framework refers to a plan that the Army would use to conduct whatever operations may be necessary in response to a terrorist attack.

[2]The Department of Homeland Security will include a Directorate of Science and Technology headed by an Under Secretary for Science and Technology. The Under Secretary will advise the Secretary on R&D efforts, priorities, goals, objectives, and policies. This might be an ideal site for the integration of civil and military research.

The Army

The Army is organized in three parts: the active Army, the Army National Guard (ARNG), and the Army Reserve. The committee believes that the ARNG will be most involved in HLS events, at least initially, because (1) it is under local (state) command, (2) it is usually closest geographically to probable sites for terrorist attacks, and (3) it is not limited in its law enforcement roles.

Equipment for the ARNG is based on its wartime mission, not its response to civil emergencies. Equipment requirements are established in the U.S. Army Training and Doctrine Command, where the ARNG has not had sufficient representation to make its needs known. Given the increased emphasis on HLS, it appeared to the committee that the ARNG should play a more significant role in determining what its HLS equipment should be.

> **Recommendation.** The Army National Guard's homeland security role must be considered in the development of the Army Science and Technology Master Plan, and resources for these requirements applied as appropriate in developing the Department of the Army Master Priority List.

Link to the Objective Force

While the Army has a long history of providing support to civil authorities, the quest for the Objective Force has great significance for the Army's future. This Army of the future is envisioned to be "more strategically responsive, deployable, agile, versatile, lethal, survivable, and sustainable across the entire spectrum of military operations from major theater war through countering terrorism to Homeland Security" (U.S. Army, 2002).

The modernization strategy that is being used to bring the Objective Force to rapid fruition envisions the acceleration of S&T (U.S. Army, 2002). While many of the Objective Force technologies are directly applicable to the Army's newly energized homeland responsibilities, it may be necessary to modify or adapt specific technologies to serve a dual purpose. In addition, some new capabilities will be needed. The committee believes that if this process is accomplished thoughtfully and flexibly, there are great opportunities for cost-effective procurements, economies of scale, and an ability to accomplish both missions successfully.

> **Recommendation.** To optimize current science and technology efforts, the Army should take advantage of potential transferability between technologies for homeland security and those for the Objective Force.

As the committee became more familiar with civilian first responder requirements, an interesting parallel began to emerge between responding to a domestic terrorist attack in close cooperation with local authorities and fighting a war in close cooperation with allies and coalitions of allies. In both situations, the Army will be working with groups who have different equipment, different cultures, different operational languages, etc. The requirement to create force packages tailored for particular incidents and to establish interoperable situational awareness and command, control, communications, computers, intelligence, surveillance, and reconnaissance (C4ISR) will be overriding.

Recommendation. The Army should investigate the technologies necessary to put together on the fly the force packages necessary to meet the requirements of both homeland security and the highly deployable Objective Force.

Recommendation. Given the time lag associated with training personnel and leadership to use new technology, now is the time to start dealing with these issues in the context of homeland security, so that they are well honed by the time the Objective Force is fielded.

INDICATIONS AND WARNING

Indications and warning (I and W) generally refers to the events leading up to an attack. Much of this is the province of the intelligence community. Since the Army will have a significant role in responding to the use of weapons of mass destruction (WMD), the committee focused in this portion of the study on the physical detection of explosives (nuclear and conventional), radioisotopes, chemical agents, and biological agents and on the identification of related cross-cutting S&T.

Traditional Imaging Sensors

The advanced, high-performance imaging systems that infuse all aspects of national security and defense also have relevance for HLS. High-performance sensors, which image in a broad range of spectral bands, are a high priority for numerous theater and national missile defense platforms. The Department of Defense (DoD) in general and the Army have broad programs in this area.

Recommendation. It is critically important that all sensors not only be well characterized at the point of purchase but also be regularly rechecked by competent technicians. Software used to integrate disparate sensors should be well documented and checked against standardized problems.

Chemical Agents

Chemical agents are typically released into the atmosphere, where they form toxic clouds that are moved by atmospheric winds or by ventilation systems. The most desirable situation would be to detect these agents before they are released into the atmosphere. For weaponized agents this will be difficult because of problems with sensitivity and false alarms when operating in realistic dirty environments.

Biological Agents

The point detection of biological agents is qualitatively different from that of chemical agents. Compared with chemical agents, many orders of magnitude less of biological agent are required to incapacitate an individual. This means that there may be substantially less material to detect. A typical biodetection system involves a cueing, detection, discrimination, and identification sequence. Unlike chemical agents, live biological agents may replicate themselves in the infected population to a detectable level, but only after their release. Replication of infectious agents in the population may also contribute to secondary spread of the disease.

Nuclear Materials

In the case of nuclear weapons, the primary fissionable isotopes of interest are uranium-235, plutonium-239, and uranium-233. In most cases detectors are effective only if they are relatively close to the source of radiation. For example, the signature from a plutonium weapon's spontaneous decay processes will be gamma rays and neutrons. Assuming scattering but no neutron capture between the weapon and the detector, the weapon neutron flux from spontaneous fission will equal the background neutron flux at about 15 meters from the weapon, making detection at a distance problematic. All of the nuclear materials detectors mentioned in the report have relatively short detection ranges and are best suited for choke points or portal geometries or where there is good intelligence on where the material is located.

Conventional Explosives

The majority of terrorist attacks against U.S. forces, facilities, and citizens have involved the use of conventional explosives. The detection and tracking of such explosives is therefore extremely important. The vapor-phase detection of a modern explosive will be possible only if there are detectors in close proximity to the explosive or if there is a very substantial concentration of explosive vapors at a distance from the explosive.

Army weapons and explosives in transit or in storage can be attractive targets for theft or diversion by terrorists. On a broader scale, it would be in the interest of the United States if international protocols were established that called for the insertion of detection markers and identification taggants, worldwide, into all legitimately manufactured explosives to assist both detection and forensic analysis.

Recommendation. An international convention requiring the incorporation of detection markers and identification taggants should be sought.

Techniques to detect packaged dangerous materials are for the most part lacking. The committee learned that such detection is an extremely difficult problem even when the detector can be placed next to the package. New and perhaps radically different approaches will be required. A distributed network could involve fixed sensors and mobile sensors deployed on various platforms including autonomous unmanned air, space, ground, and underwater vehicles. This option opens up substantial opportunities for the investment of Army S&T resources because the S&T involved is more broadly applicable to the Army than just nuclear weapons detection or chemical and biological agent detection.

Recommendation. The Army should ensure from the outset that the necessary interrelationships among the sensor networks and the broader intelligence collection activity are established and maintained as a coherent undertaking.

Recommendation. Army science and technology should aggressively seek out and invest in those cross-cutting sciences and technologies that will benefit both the Objective Force and the homeland security requirement to detect weapons of mass destruction.

DENIAL AND SURVIVABILITY

The principal element of successful denial is good security, including both physical security and cybersecurity. Denial of an attack refers to measures taken to prevent or otherwise thwart an intended terrorist attack, whether by preventing access using, for example, guards or barriers or by other means of interception (e.g., explosive detection and electronic surveillance). Survivability, in contrast, refers to measures taken to mitigate the effects of an attack by such means as structural hardening, protecting personnel, and duplicate resources. Survivability also includes the ability to absorb an attack with acceptable damage and casualties, redundancies that enable continued function after an attack, mitigation of the effects of the attack, and preparations that plan for operation afterward.

Recommendation. To gather valuable and perishable medical and other forensic data, the Army should support the establishment of rapid response data-gathering teams to investigate bombing attacks that may occur in the future. The data collected by these teams should be integrated with information from past events and made available to researchers and practitioners in emergency medicine, injury epidemiology, search and rescue, architecture, and engineering.

The fixed infrastructure targets presumed to be of primary interest to the Army are military buildings either inside an installation or standing alone (e.g., barracks, office buildings, and command-and-control (C2) centers), bridges, tunnels, and dams, as well as special facilities such as nuclear power plants and critical Department of Defense (DoD)/Army assets (e.g., ports and airfields). Infrastructure targets also can include those that are primarily "cyber"—computer networks, communication systems, and C2 systems or supervisory control and data acquisition (SCADA) systems for base power grids and water systems.

Physical Security

The technology needs for physical security are very broad. Explosive threats against conventional buildings of direct interest to the Army may range from small 1- or 2-pound explosives packaged in letter bombs or pipe bombs, to hundreds of pounds of explosives contained in cars, to thousands of pounds of TNT (trinitrotoluene) equivalent charge carried by large trucks, trains, or dockside ships.

Military and conventional buildings are susceptible to chemical, biological, and radiation attacks by terrorists through their heating, ventilation, and air-conditioning (HVAC) systems. The effectiveness of such attacks can be greatly reduced by incorporating building automation systems that can be designed to manage specific threats and scenarios.

> **Recommendation.** The Army should monitor and integrate new heat, ventilation, and air-conditioning technologies developed by the Defense Advanced Research Products Agency and other organizations into building and infrastructure design and retrofit guidelines. These technologies include detection, neutralization, filtration, and active ventilation defenses.

The Technical Support Working Group (TSWG)/Defense Threat Reduction Agency (DTRA) Blast Mitigation for Structures Program is a focused and valuable program of research, testing, engineering analysis, and computational modeling to supplement existing knowledge on blast effects and blast-resistant design and construction. However, the full benefits of the program will be realized only if the results are widely disseminated and necessary improvements implemented.

Blast-hardening technologies and design principles developed by the Army and other DoD components for military purposes are generally relevant for federal force protection and civilian design practice. However, because the knowledge base is incomplete, this information must be adapted and expanded to be more specifically usable by and accessible to civilian architects and engineers.

> **Recommendation.** The Army should continue to survey and evaluate relevant ongoing university research with the objective of identifying and synthesizing technology that could improve the performance of buildings in a

blast environment, and it should also consider inviting universities to participate directly in the research effort.

Information Security and Cyber Issues

The word "cyber" is used in this report to refer to any activities related to the computer and communications (C&C) infrastructure, including information stored and/or transmitted in the systems. Use of this infrastructure is rapidly becoming ubiquitous in all aspects of daily life. The C&C infrastructure can be compromised by several mechanisms, principally these:

- An insider making use of authorized access,
- Unauthorized access via direct tapping into the physical facility,
- Unauthorized access via valid network connections and security flaws in the system, and
- Denial-of-service attacks.

There are three primary objectives of a cyber attack:[3] (1) destroy or change data within the system itself, (2) take control of systems controlled by the C&C system, or (3) deny the user effective use of the system. Future terrorist incidents in the United States might utilize any of these. The best defense is to physically isolate an important network from the public network.

Large organizations are often tempted to custom design their own systems, because they believe their needs are different and that they can achieve greater efficiency by dropping those system elements they do not require, at least at the time of design. For general-purpose systems this is not only a false economy—the design costs are such that because of the rate of change in the field, the organization will soon be left with an out-of-date software design that runs only on out-of-date hardware—but it is also an invitation to security disasters.

Recommendation. The Army should partner with other agencies and the commercial sector to develop and adopt the appropriate tools and protocols for the protection of its own computer and communication systems.

Recommendation. The Army should continue to review its cybersecurity procedures to assure that the best practices from the community are adopted on an ongoing basis.

[3]Attacks by hackers merely to prove their abilities by making annoying but inconsequential changes to the system are not discussed. It should be recognized that many of these hacker attacks are against that part of the network that is designed to be public, that is to say public Web sites. While it is desirable to keep those pages secure against unauthorized change, the level of security that can be applied to nonpublic information is necessarily lower.

The Army must be concerned not only with the survivability of its own systems in the event of an attack but also with the survivability of systems over which it has no or little control prior to the attack—or even, perhaps, after the attack—since if it is called on to provide support, it will need to establish links between its units and civilian responders.

> **Recommendation**. Whether through the Army National Guard or active or reserve Army units, the Army should play a major role in providing emergency command, control, communications, computers, intelligence, surveillance, and reconnaissance (C4ISR) in the event of a major natural or terrorism disaster because it has both the skill set and the equipment to provide such services in hostile environments.

> **Recommendation.** Equipment and trained personnel should be available to provide vital information and communications for interoperable command, control, communications, computers, intelligence, surveillance, and reconnaissance (C4ISR) in the case that civilian systems are seriously impaired in an emergency event.

CONSEQUENCE MANAGEMENT AND RECOVERY

Generally, recovery is viewed as a local and private sector responsibility. However, in the case of terrorist acts using WMD or significant cyberattacks on the nation's critical infrastructure, the damage may exceed the capacity of local agencies and the private sector that owns and operates the critical infrastructure. Consequence management is more than just minimizing the damage; it also involves rescue of and aid to injured victims and the restoration of essential services.

Interoperable C4ISR system

The architecture and technology needed for a HLS C4ISR system is compatible with the Army's framework for developing and fielding the Objective Force. However, Objective Force C4ISR systems will need to be adapted for this different mission and different challenges.

> **Recommendation.** To facilitate the development and fielding of an integrated command-and-control system for homeland security, the Army should initiate or continue research that permits the earliest possible fielding of deployable communications packages equipped with universal multiplexer capability to facilitate C2 across the vast, and disparate, array of agencies that will respond to incidents and events.

Rapid Event Assessment of Physical Damage, Casualties, and Contamination

A necessary condition to conduct recovery and consequence management (R and CM) activities is an assessment of the situation. The Family of Integrated Operational Pictures (FIOP) is designed to meet the needs of the war fighter. However, it could be extended to the HLS mission. A number of sensors exist that can assist with a real-time situational assessment. Overhead imagery from satellites and high-endurance unmanned aerial vehicles (UAVs) can build an optical and infrared picture of physical damage. They can also use measurement and signal intelligence to determine WMD contamination. Reports and images from multiple sensors do not, by themselves, build the situational awareness and operational picture needed to conduct effective operations. The sensor pictures and reports need to be analyzed and depicted on a common grid and shared with the R and CM forces. Finally, a family of models that can predict physical damage, contamination, and casualties can play an important role in the HLS mission.

> **Recommendation.** The Army should conduct research on processes and systems to facilitate the event assessment process. It should support high-priority research such as sensor networking and fusion to merge reports from disparate sensors into a common picture.

Force Protection

The forces employed for large-scale R and CM activities need to be protected for sustained operations. Individual protection suits and inoculations are necessary to sustain operations in WMD conditions. The Army, through its Soldier and Biological Chemical Command (SBCCOM), continues to lead in the development of individual and collective protection technologies. Mobile collective protection facilities are necessary for long-term R and CM activities. The Army is currently developing a new family of deployable collective shelters that can be used by forces engaged in the HLS mission. The primary responsibility for the development of vaccines and medical countermeasures to protect against biological agents rests outside the Army in the Department of Health and Human Services and the Centers for Disease Control. However, the expertise in Army laboratories is essential to progress in this area.

> **Recommendation.** The Army's research and development across the spectrum of technologies needed for individual and collective protection against the effects of weapons of mass destruction for the Army and civilian emergency responders should be continued.

Treatment of Mass Casualties

It is likely that mass casualties will result from the use of WMD and high explosives. A mass casualty incident is one in which there are not enough resources for casualty management. In addition, triage takes on an entirely new aspect, one closely resembling the wartime rules of engagement. Where the cause of injury is suspected to be a chemical agent, toxin, or toxic industrial chemical, the responders must be able to identify the agent and determine the concentration. Methods for field assessment of biological hazards are also employed at this phase of the operation. While it is essential that the military be able to interface with civilian HLS activities as needed, some aspects of military capability may not perfectly match HLS needs.

Recommendation. The Army should expand its research in the area of triage, tracking, and treatment of mass casualties.

Recommendation. The Army should ensure development of individual triage assessment for mass casualties from events involving weapons of mass destruction.

Recommendation. The Army should ensure the development of a process to leverage information technology to effectively conduct mass casualty triage, tracking, and treatment following such an event. The process development should incorporate (1) remote decision support systems that can be integrated with civilian systems and (2) a tracking system.

Containment and Decontamination of the Effects of WMD

There is not much experience in wide-area decontamination in the aftermath of chemical, biological, and radiological/nuclear weapons attacks. Even with a correct assessment of the levels of contamination, there are few tools and techniques available for decontamination. Decontamination will probably be accomplished in stages, and it is likely that the Army will be involved in early remediation of WMD events.

Recommendation. Army science and technology should concentrate on the further development of a process to plan and implement remediation and decontamination for chemical, biological, radiological, and nuclear events. This process must be capable of being conducted in real time based on limited information.

Recommendation. Army science and technology should concentrate on the further development of decontamination solutions for chemical, biological, radiological, nuclear, or even large explosive events weapons.

ATTRIBUTION AND RETRIBUTION

In general, attribution is assigning a cause or source to an act or event. In the context of this report, it is the identification of individuals or organizations that are responsible for direct or indirect acts of terrorism and sabotage directed against the United States, its territories, and vital national interests. Retaliation is action taken in return for an injury or offense and to deter future attacks.

While the committee has no recommendations for attribution—leaving that to nonmilitary agents—the Army's role in retaliation runs the gamut from simple military/law enforcement coordination, when appropriate, to full-blown remote operations overseas, where the Army may be assigned primary ground retaliation responsibility as part of a Joint Task Force. Since this role is primary to the Army, the committee believes there are some enabling technologies that should receive very high priority and deserve S&T investment.

Operational Area and the Army Role

Operations in urban environments and in the presence of noncombatants will probably be common. The ability to move quickly in a crowded city swarming with civilians and hiding some terrorist cells is an extremely complicated task. This problem was clearly demonstrated in Somalia. The Army must be able to move personnel quickly, through or over busy streets. The committee feels that exoskeleton technology significantly increases the running and jumping capability of the individual soldier. Likewise, there is a need for small, armor-plated, light transport vehicles, ground and helicopter, to move forces as needed in this environment. Additionally, a capability is needed for clearing obstacles in the streets and alleyways.

Technology Focus Areas

One key aspect of survivability is signature reduction of our forces across the spectrum—radio frequency (RF), electro-optical, infrared, radar, acoustic, etc. Additionally, enhanced armor protection is of critical importance in the Objective Force Warrior program. Fire support plays a critical role in all combat operations. The vast majority of current fire support systems were not developed specifically for urban warfare, where precision and lethality (or nonlethality) can determine the outcome of an operation. Even relatively small errors can be devastating in terms of collateral damage or innocent civilians killed.

Recommendation. The Army should continue and enhance current research and development to focus on mobility operations in the urban environment, to include exploration of small, mobile armored carriers for use in urban environments and mini-breachers to clear streets and alleyways.

There is no good system for achieving situational awareness in an urban environment. This is due in part to the extremely complex RF propagation environment in this setting, coupled with the high-resolution accuracy needed to track a soldier in a specific room or building. A comprehensive situational awareness system building on the current Land Warrior system and linking the individual soldier to on-the-body, local, and remote sensor systems and information databases is necessary.

Recommendation. The Army should modify current systems or develop new systems, along with appropriate munitions, that are specifically designed for extremely precise fire support in urban environments.

Recommendation. The Army should make technologies such as the situational awareness Blue Force Tracking program and the health monitoring system available to the Department of Homeland Security, which will consider whether or not they can be adapted for civilian use.

Locating and tracking small terrorist cells in a rural environment is a very difficult task, particularly when the terrorist attempts to blend into the environment. Several advanced technologies may help the war fighter locate terrorists in this environment. However, there may well be a physical limitation to detector capability.

Recommendation. The Army should continue to develop a robust soldier situational awareness system begun in Land Warrior that provides a real-time, fused information system.

Recommendation. The Army should adopt a tiered approach to the problem of terrorist cell tracking and surveillance in the urban environment and in rugged terrain, first increasing sensor sensitivity, then networking and fusing sensors, and, finally, fusing information from disparate sources.

The committee believes that defense of the homeland is the military's top priority and that the Army will play a significant role in this action. Science and technology can and will assist the Army in this role.

Recommendation. The Army should focus its funding and research efforts on the high-payoff technologies shown in summary Table ES-1.

TABLE ES-1 High-Payoff Technologies

Function	Technology	Availability[a] (R, N, F)	Multiuse[b] (H, O, C)
Chapter 2	Indications and Warning Technologies		
Perimeter defense and warning	HgCdTe imaging LWIR arrays to fabricate high-performance detector arrays.[c]	R	H, O, C
	Uncooled bolometer arrays utilizing temperature-dependent dielectric constants and operating at room temperature.[c]	R, N	H, O, C
	GaAs quantum well arrays; a type of extrinsic photoconductor in which the bound electrons reside inside the quantum wells instead of on dopant ions.[c]	R, N	H, O, C
	GaN UV detectors for solar blind applications.[d]	F	H, O, C
Biological agent detection	DNA microarrays that can monitor thousands of genes simultaneously.	F	H, O, C
	Combinatorial peptides using massive libraries for screening.	F	H, O, C
	Raman scattering; matches observed Raman spectra against a library of predetermined signatures.[e]	N, F	H, O, C
Vapor-phase explosive detectors	Chemical resistors that detect at the parts per billion level. Must be close to explosive or chemical, needs improved SNR.[f,g]	N	H, O, C
	Fluorescent polymers that detect at parts per trillion level (in principle). Must be close to explosive or chemical, needs improved SNR. Demonstrated at parts per billion in reliable system.[h]	R, N	H, O, C
	Surface-enhanced Raman spectroscopy that detects at parts per billion. Portable, must be close to explosive.[h]	N, F	H, O, C
	Immunoassay (biosensors) that detects parts per billion. Must be close to explosive. Potential for increased sensitivity.[h]	N, F	H, O, C

Function	Technology	Availability[a] (R, N, F)	Multiuse[b] (H, O, C)
Bulk explosive detection	Nuclear quadrupole magnetic resonance (NQR). Low SNR, must be close to explosive, does not require magnets. Produces RF signals characteristic of particular explosives.[g,i]	R, N	H, O, C
	Millimeter-wave radiometry. Potential to provide radiometric images of objects (e.g., explosives) under clothing.[g,j]	N	H, O, C
Cross-cutting detection and tracking	Sensor networking—gathers data from a wide variety of spatially distributed sensors.	N, F	H, O, C
	Sensor fusion—intelligently combines, correlates, and interprets data from distributed sensors.	N, F	H, O, C
	Anomaly detection—examines data from networked sensors to discover patterns, unusual behavior, etc.	N, F	H, O, C
	Surveillance platforms (UAVs, UGVs, UUVs)—small autonomous vehicles for carrying sensor payloads as part of distributed sensor network.	R, F	H, O, C
Cross-cutting perimeter surveillance	IR, RF, acoustic, seismic, etc. techniques that monitor for intrusion into predetermined spaces (encampments, facilities, borders, etc.).	R, N	H, O, C
Cross-cutting capability in miniaturized systems	MEMS—methods for integration of many technologies into microsensors using electronic fabrication technologies.	R, F	H, O, C
	Active-passive sensor suites—suites of lasers and detectors that can query and image as well as perform spectroscopic measurements.	N, F	H, O, C
	Nanofabrication techniques—fabrication of sensing systems at the atomic level.	F	H, O, C

Continues

TABLE ES-1 Continued

Function	Technology	Availability[a] (R, N, F)	Multiuse[b] (H, O, C)
Chapter 3	Denial and Survivability Technologies		
Perimeter control	X-ray assessment, swimming sensors for rapid detection of LVBs.	N, F	H, O
	Unattended sensor networks, advanced power sources, C2 and secure communication, low-power sensing elements for deployable perimeter control system.	N, F	H, O
	C2 and secure communications, situational awareness tools, area sensors for mobile perimeter system.	F	H, O
Building and facility access control	Smart ID with bioinformation, ID tracking with area authorization, iris ID, liveness tests, auto DNA ID for automatic, high-confidence access control.	F	H, O, C
Structural blast resistance	Prediction of blast and impact loads on and in buildings, bridges, dams, etc.	N, F	H, O, C
	Connection details for steel and concrete structures (new and retrofit construction) to upgrade current approaches for dynamic environments and material behavior.	N	H, O, C
	Methodology to prevent/evaluate potential for progressive collapse.	N (+ university, industry)[k]	H, O, C
	Blast-resistant window concepts, including new glazing-to-frame connections.	N	H, O, C
	Blast-resistant tempered and laminated glass (stiffness, strength enhancement, ductility).	F	H, C
	First-principles analysis techniques to supplement experimental databases for design of windows and structural component retrofits.	N	H, O, C
	Software to include new test and analysis data and techniques for design and retrofit of structures in blast environments.	R, N	H, O, C

Function	Technology	Availability[a] (R, N, F)	Multiuse[b] (H, O, C)
	Integration of performance standards with building codes from a multihazard perspective.	N, F	H, O, C
Cybersecurity	IP version 6 to provide ad hoc mobile C&C networks to rapidly reconfigure systems.	N	H, O, C
	Technologies to avoid enemy intrusions, guarantee functionality.	F	H, O
	Technologies to provide alternative C&C after a disaster.	N	H, O
	IP version 6 for networks, universal radio, etc. to allow the Army systems to interoperate with other emergency services.	N	H, O
Chapter 4	Recovery and Consequence Management Technologies		
Command and control	Adaptive integrated multiplexer systems to integrate communications between multiple agencies.	N	H, O, C
	Mobile local broadband networks to pass imagery and communications.	N, F	H, C
	Blue Force Tracking to determine the location of operational personnel and assets from multiple agencies.	N, F	H, O, C
Planning	Decision support aids such as those in the Agile Commander ATD to enhance real-time planning among multiple agencies.	N	H, O
Event assessment	Family of interoperable operational pictures displays that can be shared by operational planners and implementers.	N, F	H, O, C
	Land mobile robotics that can breach obstacles to implant sensors.	R, N	H, O, C
	Sensor networking and fusion to integrate multiple sensors into a common picture.	N, F	H, O, C

Continues

TABLE ES-1 Continued

Function	Technology	Availability[a] (R, N, F)	Multiuse[b] (H, O, C)
	Real-time damage and contamination modeling to provide attack assessments based on the reports of fused sensor data.	N, F	H, O, C
Force protection	Development of improved protective mask filters and service-life indicators.	R, N	H, O, C
	Development of semipermeable membranes and self-detoxifying material for protective suits.	N	H, O, C
	Vaccine development for protection against biological agents.	N, F	H, O, C
Medical response	Chemical, biological, and radiological triage assessment cards providing C4ISR integration of data, decontamination of the patients and material, tracking of the patients, physical evidence, clothing; chain of custody.	R, N	H, O, C
	C4ISR; on-demand access to expert's network, scenario modeling/procedures to provide remote expert support for the on-site medical personnel; on-demand linkage to medical and scientific information systems, experts, and laboratories.	R, N	H, O, C
	Field-deployable diagnostic, life-support, and emergency surgical systems that can be easily and rapidly deployed; that are resistant to vibration, low environmental quality, and electromagnetic interference; and that can be operated efficiently in the presence of chemical, biological or radiological residuals.	R, N, F	H, O, C
	Field-deployable rapid-assay devices; dynamic meteorologic models of CBRN threats to provide the first responder an assessment of agents and risks for staff and patients; assessment of ongoing environmental risks.	R, N	H, O, C

Function	Technology	Availability[a] (R, N, F)	Multiuse[b] (H, O, C)
	Scenario development software based on physiologic and biochemical response to agents.	R, N	H, O
	Hemorrhage, neurological, and respiration stabilizing devices and technologies with a long shelf-life, rapid-acting agents.	R, N	H, O, C
	Vaccines and immunologic factors (including therapeutic applications), counteragents for chemical, biological, and radiological exposure with a long shelf-life, rapid-acting agents.	R, N, F	H, O
	Distributed learning platforms with AI and decision-assisting tools for CBRNE.	R, N, F	H, O
Remediation and decontamination	Development of a process to plan and implement remediation and decontamination for chemical, biological, radiological, and nuclear events.	N	H, C
	Further development and assessment of solutions to clean up chemical and biological contamination.	R, N, F	H, C
Chapter 5	Attribution and Retaliation Technologies		
Detect traffic/ activity abnormality in urban and rural locations	Multisensor fusion.	N	H, O
	Data mining techniques.	N	H, O
	Inference algorithms.	N	H, O
	Redeployable UGS.	F	H, O
Locate terror cells in areas of heavy foliage	3-D ultrasensitive lidar.	N	O
Defeat covered and concealed targets in rural environment	3-D ultrasensitive lidar.	N	O
	Multisensor fusion techniques.	N	O

Continues

TABLE ES-1 Continued

Function	Technology	Availability[a] (R, N, F)	Multiuse[b] (H, O, C)
Locate gunshots in urban environment	Ultrasensitive acoustics triangulation system.	F	H, O, C
Enhanced red force (enemy) location in urban environment	Track deconfliction algorithms.	F	O
Situational awareness	Enhanced blue force (friendly) personnel location in urban environment provided by fused GPS, RF, and dead-reckoning hardware and algorithms.	N	H, O, C
Mobility in remote urban environment	Exoskeleton for soldier platform.	F	O, C
	Light, highly survivable, signature-suppressed troop-carrying helicopter.	F	O, C
	Mobile, small-scale robotic breachers for clearing alleys, etc. in urban environment.	N, F	O, C
Remote operations	Reduced usage of signature-producing technologies.	N	H, O
	Advanced composites for lightweight armor protection.	F	H, O, C
	Advanced composites for enhanced vehicle mine protection.	F	H, O, C
	Advanced health and wound monitoring system that integrates blood pressure, heart rate, body temperature, skin penetration sensors.	N, F	H, O, C
Munitions and delivery systems designed for remote urban combat	Nonlethal munitions to include acoustic systems.	N, F	H, O, C
	PSYOP products.	N	O
	UAVs and UGVs designed for urban fire support.	N	H, O, C

Function	Technology	Availability[a] (R, N, F)	Multiuse[b] (H, O, C)
Precision insertion and targeting for warheads	Advanced propellants.	N, F	O
	Improved warhead design,	N, F	O

NOTE: AI, artificial intelligence; ATD, Advanced Technology Demonstration; CBRN, chemical, biological, radiological, and nuclear; CBRNE, chemical, biological, radiological, nuclear, and high explosive; C&C, computers and communication; C2, command and control; DARPA, Defense Advanced Research Projects Agency; EO, electro-optical; FOLPEN, foliage penetration; GPS, Global Positioning System; ID, identification; IP, Internet protocol; IR, infrared; lidar, light detection and ranging; LVB, large vehicle bomb; LWIR, long-wave infrared; MEMS, microelectromechanical systems; NSA, National Security Agency; PSYOP, psychological operations; RF, radio frequency; SNR, signal-to-noise ratio; UAV, unmanned air vehicle; UGS, unattended ground sensor; UGV, unmanned ground vehicle; UUV, unmanned underwater vehicle; UV, ultraviolet; 3-D, three-dimensional.

[a]Availability: R, ready (TRL 8-9); N, near-term (TRL 4-7); F, far-term (TRL 1-3).

[b]Multiuse: H, Army homeland security; O, Objective Force; C, civilian (first responders and others).

[c]Westervelt et al. (1991).

[d]DARPA (2002a,b).

[e]NATIBO (2001).

[f]Lewis et al. (1997).

[g]Bruschini and Gros (1997).

[h]Ward et al. (2001).

[i]U.S. Navy (2002).

[j]NRC (1996).

[k]Participation by universities and industry should be sought, because their technology, understanding, experience, and capabilities in this area are advanced, their databases are useful, and they would provide new insight and information to the program and shorten the time frame for development.

REFERENCES

Bruschini, C., and B. Gros. 1997. A Survey of Current Sensor Technology Research for the Detection of Landmines. Available online at <http//diwww.epfl.ch/lami/detec/susdemsurvey.html>. Accessed on September 24, 2002.

DARPA (Defense Advanced Research Projects Agency). 2002a. Semiconductor Ultraviolet Optical Sources (SUVOS) Available online at <http://www.darpa.mil/mto/suvos/index.html>. Accessed on October 2, 2002.

DARPA. 2002b. Solar Blind Detectors. Available online at <http://www.darpa.mil/MTO/SBD/index.html>. Accessed on October 2, 2002.

Lewis, N.S., M.C. Lonergan, E.J. Severin, B.J. Doleman, and R.H. Grubbs. 1997. Array-based vapor sensing using chemically sensitive carbon black-polymer resistors. Pp. 660-670 in Detection and Remediation Technologies for Mines and Minelike Targets II, Proceedings of SPIE, vol. 3079, A.C. Dubey and R.L. Barnard, eds. Bellingham, Wash.: The International Society for Optical Engineering.

NATIBO (North American Technology and Industrial Base Organization). 2001. Biological Detection System Technologies Technology and Industrial Base Study, February, Available online at < http://www.dtic.mil/natibo/>. Accessed on September 23, 2002.

NRC (National Research Council). 1996. Airline Passenger Security Screening: New Technologies and Implementation Issues. Washington, D.C.: National Academies Press.

U.S. Army. 2002. Weapon Systems 2002. Washington, D.C.: Government Printing Office.

U.S. Navy. 2002. Department of the Navy Explosive Detection Equipment-Explosives. Available online at <http://explosivedetection.nfsec.navy.mil/explosives./htm>. Accessed on September 24, 2002.

Ward, K.B., A. Ervin, J.R. Deschamps, and A.W. Kusterbeck. 2001. Force Protection: Explosives Detection Experts Workshop, NRL/MR-MM/6900—01-8564, CDROM. Arlington, Va.: Office of Naval Research.

Westervelt, R., J. Sullivan, and N. Lewis. 1991. Imaging Infra-red Detectors. JASON report number JSR-91-600. McLean, Va.: Mitre Corporation.

1

U.S. Army Role in Homeland Security

INTRODUCTION

The nation's military, particularly the Army, has a long tradition of providing assistance to local, state, and federal agencies in mitigating the effects of manmade and natural disasters; providing for the public safety; and restoring essential services. In the 21st century, the scope of this mission will increase in response to the new threats and challenges. The possibility of terrorists using chemical, biological, radiological, nuclear, or high explosive (CBRNE) weapons has placed dramatic new responsibilities on the civilian emergency responder community and the military. Additionally, the potential for adverse effects of cyberattacks on the nation's critical infrastructure has increased as the infrastructure enhances its dependence on internetted communications and digital control systems.

In response to the threats, challenges, and missions, the President has signed into law a cabinet-level department, the Department of Homeland Security (DHS). The Department of Defense (DoD) has created the U.S. Northern Command (NORTHCOM), which, among other things, will organize and employ the assets of the military when it becomes necessary to meet these challenges. The Army will play a role in the new organizational structure and in meeting the new organizational demands, but the exact role remains in a general state of flux.

It is clear that the first to respond[1] to terrorist events will be the local civilian emergency responders, such as policemen and firemen. However, in

[1]Appendix D provides an excerpt of the Federal Response Plan (FRP) that outlines how the federal government implements the Robert T. Stafford Disaster Relief and Emergency Assistance

events of national significance that exceed the capabilities of the state and local authorities, the Army will most likely be called upon to assist the lead federal agency, the DHS, and the Army National Guard (ARNG) will most likely be the first Army component to assist in assuring public order, mitigating the effects of the terrorist events, and beginning the recovery for both the public and private sectors.

Finding 1-1. Homeland security is an important extension of the Army's historical role of providing military support to civilian authorities. The Army will be called on to assist the lead federal agency, the Department of Homeland Security, in meeting a wide range of demands for consequence management and recovery of public order and critical services.

For the purposes of this report the committee is using the definitions in Box 1-1, obtained from the *Department of Defense Dictionary of Military and Associated Terms* (DoD, 2001).

ORGANIZATION OF THE ARMY

It is important for the reader to understand how the Army is organized and the current limitations of the Posse Comitatus Act before considering how the Army might assist civil authorities.

Organization

While every military unit in the Army is organized, trained, and equipped using a single Army standard, the U.S. Army has three distinct components:

- The active Army,
- The Army National Guard (ARNG), and
- The Army Reserve (USAR).

The active Army is immediately available for use in an emergency. It has a balance of combat, combat support, and combat service support forces. The ARNG (with mostly combat units) and the USAR (with mostly combat support and service support units) are in reserve status and generally require a period of

---

Act to assist state and local governments when a major disaster or emergency overwhelms their ability to respond. Both the FRP (Federal Emergency Management Agency. 1999. Federal Response Plan, 9230.1-PL, April, available online at <http://www.fema.gov/rrr/frp/> and accessed on December 3, 2002) and the Terrorism Incident Annex (Federal Emergency Management Agency. 1999. Terrorism Incident Annex, April. Available online at <http://www.fema.gov/rrr/frp/frpterr.shtm> and accessed on December 3, 2002) are important for planning purposes.

BOX 1-1
Definitions

Terrorism—the calculated use of unlawful violence or threat of unlawful violence to inculcate fear; intended to coerce or to intimidate governments or societies in the pursuit of goals that are generally political, religious, or ideological.

Antiterrorism—defensive measures used to reduce the vulnerability of individuals and property to terrorist acts, to include limited response and containment by local military forces.

Counterterrorism—offensive measures taken to prevent, deter, and respond to terrorism.

time to activate before they become available. This period ranges widely, from a few hours to several weeks. The active Army and the USAR are considered federal forces and are at the call of the President; the ARNG has a dual status that allows state governors to use these forces without having to call on the federal government.

Posse Comitatus Act

The Army's role in the United States is circumscribed by the Posse Comitatus Act of 1878 (PCA) (18 USC 1385), which, as a general matter, prevents the Army (and, by extension, the Air Force) from directly engaging in law enforcement activities such as search, seizure, arrest, and similar actions.[2]

PCA applies unless the Congress has specifically authorized such direct law enforcement actions by other statute or unless the emergency is of such significance that the President may exercise his direct executive authority under the Constitution.[3]

The PCA applies to Army active forces, to members of the USAR serving on active duty or active duty for training, and to the ARNG when in federal status. Importantly, the ARNG not in federal status serves at the direction of a state governor and may perform law enforcement functions consonant with the laws of the state. It should be pointed out, however, that active forces may protect federal property necessary to the performance of a federal function and may always act in self-defense and thus may be involved in the protection of critical infrastructure and in force protection roles in the United States.

[2]While the Navy and the Marine Corps are not constrained by the PCA, similar restrictions are imposed by DoD policy.

[3]There are many statutory authorizations for use of the military in a law enforcement role under circumstances specified in those statutes. The earliest and broadest authorizing statute is the Insurrection Act (10 USC 331 et seq.), which has been used a number of times in history.

HOMELAND SECURITY

The concept of homeland security (HLS), while certainly not new, has not received a high priority within the nation until now. As this report is being prepared, there is no certainty about how the newly established NORTHCOM would proceed with the military role, and no specific role has been assigned to the Army. However, there is no doubt that NORTHCOM will have requirements for HLS that the Army and the other military services must meet. To the extent that these requirements cannot be satisfied with current resources, they will help shape the Army's Science and Technology Master Plan. The committee will consider future Army science and technology (S&T) requirements, as driven by NORTHCOM, in later studies, as the requirements evolve.

The committee proceeded on the assumption that the Army will play a significant role in HLS. A statement by Secretary of Defense Donald Rumsfeld reinforces this assumption:

> With regard to supporting the effort to improve security at home, there are three circumstances under which DoD would be involved in activity within the United States.
>
> > Under extraordinary circumstances that require DoD to execute its traditional military missions. . .
> >
> > In emergency circumstances of a catastrophic nature. . .
> >
> > Missions or assignments that are limited in scope where other agencies have the lead from the outset (Rumsfeld, 2002).

Army Homeland Security Operational Framework

In anticipation of future taskings, the Army has developed a notional operational framework[4] for HLS,[5] consistent with the National Security Strategy,

---

[4]Operational framework refers to a plan that the Army would use to conduct whatever operation may be necessary in response to a terrorist attack

[5]The following definitions are provided from Greg Bozek, Army War Plans Division, Army Deputy Chief of Staff, G3, briefing to the committee on May 15, 2002:

- *Homeland Security*: The preparation for, prevention of, deterrence of, preemption of, defense against, and response to threats and aggressions directed towards U.S. territory, sovereignty, domestic population, and infrastructure; as well as crisis management, consequence management, and other domestic civil support.
- *Homeland Defense*: The protection of U.S. territory, sovereignty, domestic population, and critical infrastructure against external threats and aggression.
- *Civil Support*: Department of Defense support to U.S. civil authorities for domestic emergencies and for designated law enforcement and other activities.

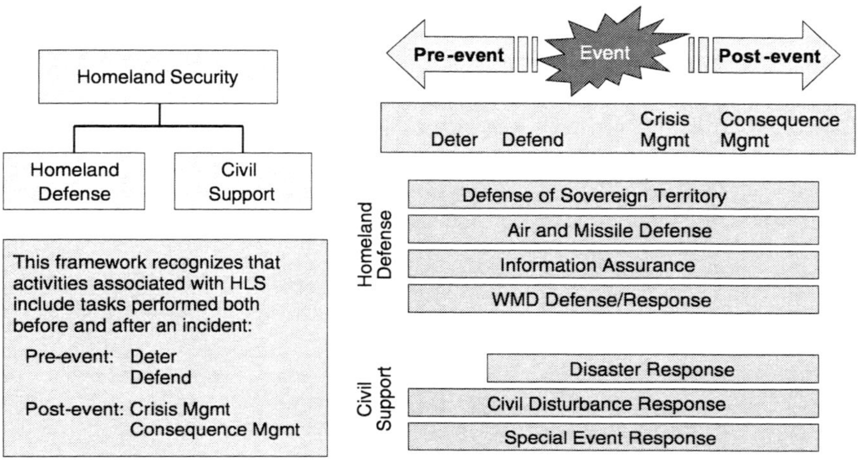

FIGURE 1-1 Army homeland security operational framework. SOURCE: Greg Bozek, Army War Plans Division, Army Deputy Chief of Staff, G3, briefing to the committee on May 15, 2002.

under which it includes "homeland defense" and "civil support." This is illustrated in Figure 1-1.

These missions extend across a continuum from pre-event to event to post-event and incorporate sequential tasks—that is, deterrence, defense, crisis management, and consequence management.[6] The President's announcement at the 2002 United States Military Academy commencement ceremony regarding preemptory strikes indicates that the model is still a work in progress.

The committee views the missions of "defense of sovereign territory" as a traditional military warfighting mission and beyond its purview, although there are surely many technological developments that would be of interest. The challenge of "air and missile defense" has already been the topic of several excellent studies. Given the limited time frame for this effort, the committee has chosen not to address this element in the report. "Information assurance" and "weapons of mass destruction (WMD) defense/response," while not new, surely pose new challenges and probably involve new functions or new applications of existing technology.

[6]Greg Bozek, Army War Plans Division, Army Deputy Chief of Staff, G3, briefing to the committee on May 15, 2002.

Homeland Defense

While perhaps least likely, scenarios exist that would demand a Presidential determination and the use of traditional military capabilities in the United States. Conceptually such an event could create the need for special capabilities related to operating in urban terrain but intermingled with and surrounded by U.S. civilians. The major capabilities required to deal with a substantial combatant force or for law and order could involve all three Army components. Additionally, as pointed out in the Army After Next Summer 1997 War Games and subsequent war games such as the Ellipse Series, conducted by Joint Forces Command, the threat to the homeland could limit the national command authority's flexibility in overseas operations owing to concerns about protecting the homeland (Brennan, 2002).

Civil Support

As previously indicated, the Army has always been available for support of civil authorities. Civil support missions, while long a part of Army responsibilities, take on greater significance in a terrorist environment. The increased sophistication and capability of the terrorist threat require planning for events of catastrophic proportions. Circumstances involving the leveling of significant portions of cities and/or the use of WMD, with perhaps hundreds of thousands of casualties, could require the Army to assist in ways never before anticipated, both in support of civilian emergency responders and as an emergency responder on military installations. The size of the requirement for DoD support to special events (e.g., the Olympic Games and the Super Bowl) has also increased many times over, both in the number of events to be protected and the variety of functions to be performed.

The Army does have substantial relevant capability within its three components.[7] However, the ARNG, in its state role, will most likely be the second responder on the scene after civilian emergency responders. The ARNG has organic units such as medical, heavy equipment engineer, military police, and communications units with inherent mobility and self-sustaining capability. It also maintains WMD response units, called civil support teams (CSTs), in many states. Depending on the situation, the active Army and the USAR will be available to reinforce the ARNG with whatever capability is necessary. The active Army has some unique capabilities when it comes to the detection of

[7]It must be acknowledged that members of the emergency first responder community, such as firemen, police, and emergency medical personnel, may also be serving in the Army's reserve components. DoD directives and Army regulations address such conflicts, requiring that they be resolved in peacetime by screening out of the Army reserves those individuals whose employers will not agree to release them for military duty.

biological and chemical weapons and consequence management of the aftermath of their use. These include some important and unique laboratory capabilities, large-scale decontamination units, explosive ordnance disposal teams, and medical research. The medical corps and laboratories also have considerable capabilities in forensic medicine and wound ballistics. The Army Corps of Engineers (USACE) has considerable capabilities that have been used frequently in support roles. Likewise, other major Army activities such as the Army Materiel Command (AMC) can be called upon to provide equipment, supplies, and other assistance to civilian agencies.

> **Finding 1-2.** The Army National Guard, given its historical mission and flexibility, geographic dispersion, dual-mission capabilities, and frequent association with local agencies, is the key Army asset to meet homeland security demands and can be augmented as necessary with special capabilities from the Army Reserve and the active Army.

Organizational Vacuum

While it is not the task of the committee to make recommendations on organizations and functions, the apparent absence of an adequate and integrated national structure for the prevention of terrorist actions and intelligence sharing from the federal to the local (first response) level was striking.

To help it analyze the need for science and technology, the committee made use of a suggested HLS concept of operation and roadmap (see Box 1-2). It should be emphasized that this is just one example of how preparations might be made and information might be shared.

The Army's Role

The committee has come to believe that the roles of the Army in HLS and in traditional war fighting, while quite distinct, will share certain similarities. One way to look at the Army's potential role in HLS is to view it in terms of the five major functions the Army must accomplish. HLS operations modify but do not drastically change these fundamental functions:

- Protecting the force,
- Projecting the force,
- Conducting operations,
- Sustaining the force, and
- Redeploying the force.

Protecting the force continues to be an umbrella concept that cuts across all five functional areas. It is continuous in nature in that it starts at a home installa-

BOX 1-2
Notional Homeland Security Roadmap

Presidential Decision Directive 63 (PDD 63) or an updated version of it will serve as a base document for federal, state, and local efforts.

1. *Conduct a vulnerability assessment.* Each sector leader conducts a vulnerability assessment of his or her sector. That assessment should result in infrastructure being classified and prioritized according to criticality.

2. *Establish priorities.* Develop a common definition for each priority level. Once the definitions are agreed upon, commence simultaneous assessment efforts at the state and federal level.

3. *Use established points of contact.* Use points of contact already established by the input of each state to the Office of Domestic Preparedness in response to requirements set forth in the Fiscal Year 1999 State Domestic Preparedness Equipment Program.

4. *Integrate prioritized lists.* After sector assessments have been completed, integrate them into one prioritized list, at both the state and federal levels. The federal government should only protect infrastructure that services multiple states or regions or is critical to national security.

5. *Share critical information.* Once the Secretary for Homeland Security has approved the federal list, it should be shared on a close-hold basis with each state. The state should then ensure completeness and deconflict any duplication at the state or local level.

6. *Assign responsibility.* The National Guard should be given the mission of developing and implementing plans to protect all federal critical infrastructures and should take the lead for the military in the overall effort of combating terrorism on U.S. soil, including assisting with the training of civilian emergency responders.

7. *Partner with private industry.* A coordinated plan to protect all state infrastructures must be developed by each state. It should fully leverage public and private efforts at the state level.

8. *Address the resource issue.* Bands of preparedness need to be developed, with the highest band of preparedness being "resource unconstrained." Realistically that will have to be modified, but not until the best level of protection that can be provided our citizens has been determined. The minimum level of protection should be that associated with protecting all critical infrastructures.

9. *Provide for minimum-level protection as soon as possible.* The resources required to provide the minimum level of protection should be established and provided through a combination of federal, state, and private funding as quickly as possible.

10. *Institutionalize the effort.* Regional Centers of Excellence need to be established across the United States to provide an independent assessment of regional plans and a means of improving identified weaknesses in areas of first responder training, equipping, and technology.

11. *Execute the long-range plan.* A long-range plan must be developed, reviewed annually, and updated as required. The goal of the long-range plan should be to provide the optimum level of security for each state and its citizens over time.

tion prior to deployment and closes back on itself with redeployment. In many ways a military installation can be viewed as another element of critical infrastructure and subject to the same considerations (threat analysis, early warning, increased security, etc.) as other critical infrastructures located throughout the United States and overseas. This remains a responsibility 24 hours a day and 7 days a week.

Projecting the force means that generally the force will have to move from a home base to conduct operations. Protection of vital lines of communication (air, land, and sea) must be assured. HLS has added a new dimension to the "fort to port" challenge. The movement of military units to the site of a domestic terrorist attack will be just as challenging—if not more so—than their simple movement to a port of embarkation.

The *conduct of operations* will generally involve operating with other organizations. These may be allies, coalition partners or, maybe, emergency first responders. Operations bring all the issues associated with compatibility and commonality. For an already manpower-constrained, capabilities-based force, the additional requirements associated with HLS make it even more imperative that S&T be leveraged to the extent possible to free up much-needed manpower. Unlike traditional operations, HLS missions will be accomplished in cooperation with a wide range of civilian local, state, and federal agencies. In considering the challenges of this new role, the committee was struck by the similarities between the Army's new requirement for cooperative work with U.S. civilian emergency response agencies and the existing requirement for cooperative work with allies and various coalition partnerships. The requirements for interoperability, particularly in communications, are identical.

Without maximizing technology, *sustaining* the force can put an unacceptable burden on our limited lift and other logistical assets.

Finally, as we have learned so often, the Army must *redeploy* to its home base, recover, and prepare to do it all over again—and they must do it throughout the period of great vulnerability.

> **Finding 1-3.** There are many similarities between military operations involving allied or coalition forces and operations involving civilian emergency responders.

Link to the Objective Force

The Army has been rethinking its concepts of warfighting since the end of the Cold War. The Objective Force is the Army's future full-spectrum force that planners envision to be "more strategically responsive, deployable, agile, versatile, lethal, survivable, and sustainable across the entire spectrum of military operations from major theater war through countering terrorism to Homeland Security" (U.S. Army, 2002).

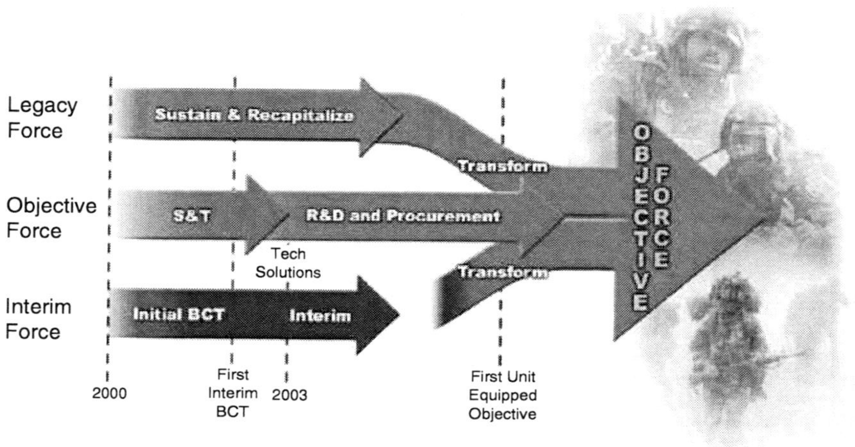

FIGURE 1-2 Army transformation. SOURCE: Andrews (2002).

Even as the Army continues its transformation into the post-Cold-War force, today's soldiers, frequently called the "legacy force," must be fully prepared to execute their responsibility to fight and win decisively against any enemy. The final transformation to the Objective Force, which begins in fiscal year 2008, will require many years of effort. The legacy force, in the meantime, will require sustainment and recapitalization to enhance its existing weapons, systems, and platforms to increasingly provide some of the Objective Force-like capabilities. As a transition from the legacy force and a vanguard for the Objective Force, an "interim force" fills the near-term gap. The interim brigade combat team (BCT) is a key element of the interim force. Figure 1-2 depicts the synergy between these forces.

The modernization strategy that is being used to bring the Objective Force to fruition envisions the acceleration of science and technology (U.S. Army, 2002). The committee believes that many of the requirements for HLS can provide a direct link to the capabilities-based Objective Force.

As indicated in the preface, the Army sponsor chose to use DoD's Defense Counter-Terrorism Technology Task Force taxonomy as a framework. This taxonomy involves the following operational areas (DoD, In press):

- Indications and warning,
- Denial and survivability,
- Recovery and consequence management, and
- Attribution and retaliation.

Using this taxonomy, the following linkages between the Objective Force and HLS are made:

- Indications and warnings technologies directly relate to protecting the force. Regardless of whether one is protecting an installation or conducting combat operations, there is a vital need for intelligence. The means of obtaining the intelligence and early warning (IEW) might vary, but the requirement is the same whether protecting an installation or conducting a combat operation. This IEW must then be transformed into actions that protect soldiers, units, and installations.
- Denial and survivability are combat multipliers for all operations as well as subsets of force protection. Leaders responsible for conducting the wide range of missions associated with today's Army must optimize the use of manpower and technology to ensure survivability.
- Recovery and consequence management can be equated to conducting combat operations and redeployment. Assisting civilian emergency responders with the consequence management of a manmade disaster could be viewed as not much different than conducting coalition operations. Recovery is a subtask associated with redeployment.
- Attribution and retaliation encompass a special form of combat operations. The Army must make available trained and ready forces to determine accountability and hold accountable the perpetrators of terrorism. The Army will most generally find itself supporting the civilian emergency responders or the combat commander, depending upon whether the terrorism is domestic or international.

Conclusion 1-1. Many of the technological requirements for homeland security will be important for the Objective Force.

Recommendation 1-1. To optimize current science and technology efforts, the Army should take advantage of potential transferability between technologies for homeland security and those for the Objective Force.

A great deal of effort and considerable resources are being directed at HLS in the civil sector. Much of the S&T effort of great interest to the Army is being conducted by agencies outside the Army. This commercially developed equipment might have great applicability for the Army, but there does not appear to be a mechanism for integrating the research being done in the civilian community with that being done in the military community. It should also be recognized that technology transfer to the civilian sector will be necessary in order for the civilian sector to exploit Army technology, and this technology transfer should be viewed as an integral element of the Army's HLS mission. The committee could not identify an integrating process whereby a single agency[8] was aware of all of this

[8]The Department of Homeland Security will include a Directorate of Science and Technology headed by an Under Secretary for Science and Technology. The Under Secretary will advise the Secretary on R&D efforts, priorities, goals, objectives, and policies. This might be an ideal site for integration of military and civilian research.

activity; therefore, no one is certain of all that is being done. This makes it very difficult for the Army to conduct a gap analysis.

Conclusion 1-2. There needs to be better means to coordinate the homeland security science and technology efforts of the Department of Defense and those of the various civilian agencies.

Recommendation 1-2. The Army should encourage better coordination of the disparate homeland security science and technology efforts.

Conclusion 1-3. Homeland security technologies developed by the Army could be of great benefit to the private sector and to other government agencies.

Recommendation 1-3. The Army should facilitate technology transfer in order to allow the private sector and other government agencies to exploit the homeland security technologies it develops.

Experience over the last decade has taught us that the use of military forces in these situations will require a tailored force package. That is to say, certain types of military units will have to be used together in a coordinated fashion. Most of the Army's experience in this area has been gained without the pressure of time. The Mission Rehearsal Exercise Model has served the Army well. However, September 11, 2001, shattered that model and has forced us to think about *no-notice adaptive force packaging*. It is the committee's belief that this will become the norm rather than the exception to the rule.

Conclusion 1-4. The ability to rapidly deploy a capability-based task force in support of either the homeland security mission or an Objective Force mission will become even more critical.

Recommendation 1-4a. The Army should investigate the technologies necessary to put together on the fly the force packages necessary to meet the requirements of both homeland security and the highly deployable Objective Force.

Recommendation 1-4b. Given the time lag associated with training personnel and leadership to use new technology, now is the time to start dealing with these issues in the context of homeland security, so that they are well honed by the time the Objective Force is fielded.

By having preplanned task forces available, the Army will not only provide better assistance to civilian emergency responders but also will be able to perfect the required techniques by the time the technology associated with the Objective

Force is fielded. The committee draws attention to this implied task to give an example of the synergy between HLS and Objective Force missions.

RESEARCH AND DEVELOPMENT FOR THE ARMY

The Army has traditionally taken care of its own needs for new technology. In the earliest days, hardware was largely built at its own arsenals, but this is largely done now by a broad set of R&D players. Most equipment is manufactured in the private sector, some by contractors and some by commercial suppliers, from whom it is procured off-the-shelf. The R&D for specifying these purchases is done in Army and other military laboratories, in academic laboratories sponsored by the Army, and in industrial laboratories under Army contracts. In recent years alliances of the Army, academe, and industry have been formed to improve the Army's focus in key areas.

The management of R&D is assigned to different departments, as well as to offices and entities at DoD. The AMC manages most of the R&D through the Army Research Laboratory and the AMC major subcommands that focus on product areas. Each of these has a Research, Development, and Engineering Center that performs 6.2 and 6.3 work—some in-house, but most under contract.[9] Separately, the Army's Medical Commands and the USACE perform their own R&D. Additionally, the USACE Engineer Research and Development Center has the DoD-wide S&T (6.1-6.3) lead for Survivability and Protective Structure for explosive threats. Within the Army the USACE center has a force protection mission, but other participants are responsible for various aspects of the CBRNE threat spectrum. The Deputy Assistant Secretary of the Army for Research and Technology supervises the 6.1-6.3 work for nearly all of AMC and for some of the medical and USACE work. A separate office in DoD oversees chemical and biological warfare work for which the Army is the principal executing agent through the U.S. Army Soldier and Biological Chemical Command (SBCCOM) and the various medical commands, such as the Medical Research and Materiel Command. The Defense Advanced Research Projects Agency (DARPA) carries out work for and with the Army. The most notable DARPA program today is the Future Combat Systems program. Being well aware of the divided R&D responsibilities, the committee decided to review as broad a range as possible of technologies that might be of help to the Army for HLS; it did so for completeness and to make the report more valuable.

The committee believes it is important to recognize an *Army* challenge to the S&T problem associated with HLS. From everything the committee has heard, it is clear that the ARNG will play a significant role for the Army in HLS. Any

---

[9]Basic research, 6.1; applied research, 6.2; advanced technology development, 6.3 (U.S. Army, 2001).

definition of the critical requirements associated with HLS must have ARNG input, and appropriate resources must be applied to the Department of Army Master Priority List. HLS is an area in which the active Army will find itself supporting the ARNG, and this must be recognized when developing the Army Science and Technology Master Plan. The committee feels that the Army must make that commitment to the ARNG.

Conclusion 1-5. The Army National Guard does not appear to play a direct role in defining the critical requirements associated with homeland security.

Recommendation 1-5. The Army National Guard homeland security role must be considered in the development of the Army Science and Technology Master Plan, and resources for these requirements applied as appropriate in developing the Department of the Army Master Priority List.

SCENARIOS

Threats are classified in terms of scenarios, the most common being CBRNE. Scenarios vary widely in their effects. We usually think of nuclear scenarios as very large and regional in effect. Explosive/incendiary scenarios are more likely to be limited in area of impact. The Pentagon disaster, for example, affected only a single building.

Our built environment has been constructed largely on the basis of ordinary occurrences and does not consider warlike disasters. Thus the usual building codes do not provide resistance to bombs, firestorms, nuclear blasts, and the like. There are exceptions. Resistance to earthquakes is now required in many seismically active regions. Wind resistance is specified in certain buildings. And we find that the World Trade Center was designed to resist the impact of an aircraft but not the combination of that plus a conflagration. Nor have we designed our environment with chemical or biological attacks in mind. So we are now faced with having to rethink these hazards and plan our response to them.

One can imagine a wide range of effects for chemical, biological, nuclear, and radiological (dirty bomb) attacks. The committee reviewed scenarios created by the following organizations: Rand Corporation, Hicks & Associates (SAIC), DARPA, and the Office of the Deputy Assistant Secretary of the Army for Research and Technology.

Funding responsibilities are divided within the government, and the Assistant Secretary of the Army for Acquisition, Logistics, and Technology is not responsible for planning investment strategies for chemical, biological, and nuclear threats. Nonetheless, the committee felt that it had to consider all scenarios to ensure the completeness of its work. Furthermore, it needed this

BOX 1-3
Some Sample Scenarios

Chemical[1]	*Biological*[1]	*Nuclear*[1]
Tank cars of phosgene are blown up; the liquid vaporizes and spreads over a city.[2]	Anthrax or smallpox is spread over a city by crop duster.[2]	Explosive devices.
Radiological	*Explosive/incendiary*	*Cyber*
Radioactive material is spread by conventional explosive on the ground or by a small plane.	Truck bombs; suicide bombers; aircraft as projectiles.	Disruption of the following via cyberattack: DoD command and control systems Power distribution (SCADA) Air traffic control systems Public switched network control

[1]The committee is aware that the Office of the Deputy Assistant Secretary of the Army for Research and Technology is not responsible for funding R&D for chemical, biological, or nuclear scenarios. Nonetheless, combating terrorism requires an integrated approach to all threats. Some of the technologies will apply to all or most threats and thus must be considered together by the Army operators.

[2]Dennis VanDerlaske, Office of the Assistant Secretary of the Army for Acquisitions, Logistics, and Technology, briefing to the committee on May 14, 2002.

better understanding to consider the additional priorities that should be established when a particular technology applies to more than one scenario. The scope and nature of the scenarios tend to dictate both who the principal responders are and the functional capabilities required for dealing with the scenario. Thus, local civilian emergency responders will probably deal with a limited conventional attack, with support from other governmental levels as necessary. A major regional disaster will certainly require state-level response, including the ARNG. In the more severe cases, full military support to civilian authorities will be required, and there may have to be Presidential declarations enabling the full use of federal resources, including federalizing the ARNG and using the active and reserve Army components to maintain law and order.

For the present study these distinctions are important, as each type of scenario is considered, particularly in recovery and consequence management. See Box 1-3 for some scenario examples. Note that in most localized disasters certain elements of the active Army are called upon to assist, but, as indicated above, not for law enforcement. Such elements might include USACE, the various Army medical commands, and the logistical capability of the AMC. In a chemical or biological attack, the AMC's SBCCOM will certainly be involved. In developing the capabilities required to address each of the four operational areas—indications and warning, denial and survivability, recovery and consequence management, and attribution and retaliation—the committee had to determine those capabilities that apply to all scenarios and those that are required only for certain scenarios. Thus some scenarios will require massive evacuations; others, vaccinations and quarantines.

FUNCTIONAL CAPABILITIES AND ASSOCIATED TECHNOLOGIES

The committee's working groups determined that certain functional capabilities were required in each of the operational areas. These, in turn, were broken down into technologies. It is important to note that for this overview, the committee decided to review technical areas one level above a particular project, device, or technique. For example, the committee looked at uncooled infrared night vision as a field capability but not at specific devices presently under development. Once the committee had determined the individual technical areas, the assessments began.

This broad survey of relevant technologies was undertaken to gain some general understanding of the S&T involved. The assessments are meant to assist in devising an R&D investment roadmap for the Army. The committee studied each technical area enough to judge its maturity. The study comments on the state of the technology and on the appropriateness of funding work in this area with Army S&T funds. In some cases it was found that S&T that was important to both the Army Objective Force and to HLS did not warrant the expenditure of Army S&T funding because other agencies or organizations have principal responsibility for that area and S&T funds are appropriated for the other agencies. Regardless of which agency is responsible for the S&T, the Army must be appropriately equipped with the products of this S&T if the products affect the missions assigned to the Army.

Each chapter contains a series of tables that display the committee process. The committee offers its judgments on priority by functionality, technology, characteristics, availability, priority for Army S&T, and uses. Each of the tables is somewhat different, depending on the information it was intended to present; however, all will follow the same general outline.

Functionality is the broadest parameter and is intended to describe, in a general sense, what the technology should be able to do. The technology column is technology at a generic level, not to be construed as suggesting a specific system. The characteristics column provides a general description. Availability is described by clustering the technology readiness levels (TRLs) into three groups. A full description of TRLs is available in Appendix C. Throughout these tables, the following code is used to describe availability:

- R, ready (TRL 8-9);
- N, near (TRL 4-7); and
- F, far (TRL 1-3).

The committee next gives its opinion on priorities for S&T investment. The following gradients are used:

- Low, someone else has the mission or the technology is ready and available;
- Medium, useful but of limited impact and some investment is needed; and
- High, very important, no one else is working on it, and considerable investment is needed.

In some cases parenthetical entries suggest that participation by universities and/or industry should be especially sought because their technology, understanding, experience, and/or scientific capabilities in these areas are advanced, their databases are useful, and their participation would provide new insight and/or information to the program and shorten the time frame for development. A summary table appears at the end of Chapter 6 that displays each of the technologies that the committee has rated high—that is, very important for Army S&T. This table will provide insights for high-payoff technologies.

As mentioned above (see the section "Scenarios"), some of the technologies will apply to more than one scenario or be used by more than one type of operator. The committee termed these multiuse technologies, and it tended to give them a higher priority. The following code is used throughout the report tables:

- H, homeland security (HLS);
- O, Objective Force (OF); and
- C, civilian.

As the committee process progressed, it became clear that certain technologies were of such universal significance that they crossed operational boundaries. The committee feels that command and control, communications, computers and intelligence, surveillance and reconnaissance (C4ISR) will be of supreme importance and will apply to a greater or lesser extent in each of the four operational areas. Therefore C4ISR is an implied subarea in each task. Similarly, medical response is a major component of crisis response and must be a part of the discussion on recovery and consequence management.

Conclusion 1-6. Command, control, communications, computers and intelligence, surveillance, and reconnaissance (C4ISR) is of supreme importance and will apply to a greater or lesser extent in each of the four operational areas in both homeland security and the Objective Force.

SUMMARY

In conclusion, the committee believes that the S&T required by the Army for HLS need not be unique. The S&T work already being done for the Objective Force could provide much of the technology needed for HLS. In fact, if approached properly, the HLS effort not only can help to advance the S&T needed for the Objective Force, but can also assist in developing tactics, techniques, and procedures. The ARNG is critical to the success of the Army's efforts in HLS. In order for the Army to successfully meet the challenge of the HLS mission, all components of the Army must work together.

REFERENCES

Andrews, A.M. 2002. Army Science and Technology...Accelerating the Pace of Transformation. Briefing by A. Michael Andrews, Deputy Assistant Secretary of the Army for Research and Technology, to the Committee on Review of the Effectiveness of Air Force Science and Technology Program Changes. The National Academies, Washington, D.C., August 22.

Brennan, R. 2002. Protecting the Homeland: Insights from Army Wargames. Available online at <http://www.rand.org/publications/MR/MR1490/MR1490.pdf>. Accessed on October 3, 2002.

Department of Defense (DoD). 2001. Department of Defense Dictionary of Military and Associated Terms. Available online at <http://www.dtic.mil/doctrine/jel/doddict>. Accessed on September 6, 2002.

DoD. In press. Defense Counter-Terrorism Technology Task Force. Alexandria, Va.: Defense Threat Reduction Agency.

Rumsfeld, D. 2002. Testimony of the Secretary of Defense before the United States Senate Committee on Appropriations, May 7.

U.S. Army. 2001. Army Science and Technology Master Plan. Washington, D.C.: U.S. Army, Office of the Deputy Assistant Secretary of Defense for Research and Technology (Chief Scientist).

U.S. Army. 2002. Weapon Systems 2002. Washington, D.C.: Government Printing Office.

2

Indications and Warning Technologies

INTRODUCTION

Within the Army homeland security (HLS) framework, indications and warning (I and W) would be classed as a pre-event undertaking. I and W generally refers to the ability to detect events leading up to an attack. These events might involve enemy planning of the attack, its identification of targets of the attack, its acquisition of materials needed for the attack, its positioning of materials to carry out the attack, and, finally, its launch of the attack itself. Much of this is the province of the intelligence community within the civilian sector rather than of the Deputy Assistant Secretary of the Army for Science and Technology (DASA (S&T)). However, as materiel is moved so as to become an imminent threat to the Army, the science and technology (S&T) necessary to allow the Army to detect the presence of this material or its movement is the legitimate responsibility of DASA (R&T), as is the S&T associated with detecting the launch of the attack itself. In some cases the S&T resources necessary to meet these responsibilities reside with other agencies.

Since the Army will have a significant role in responding to any use of weapons of mass destruction (WMD),[1] this study focused on the physical detection of or the movement of explosives (nuclear and conventional), radioisotopes, chemical agents, and/or biological agents and the identification of related S&T, which is cross-cutting in character. The Army is responsible for defending its own forces at home and abroad and will need to acquire the technology to do so

[1]The important topic of I and W in cyberspace was not addressed due to the short duration of the study.

regardless of which agency is responsible for the development of that technology. For this reason, in the summary charts that follow some areas that are designated as very important to the Objective Force and to HLS are nonetheless assigned a low priority for the use of Army S&T funds. This does not necessarily mean that the expenditure of S&T funds for these areas has a low priority; rather, it often means that organizations other than the Army are responsible for the required S&T investments.[2] The committee found cross-cutting technologies[3] such as the networking of distributed sensors, data fusion and advanced materials that are strong contenders for Army S&T and that would also be of great benefit to the WMD detection problem.

The traditional I and W for threats to Army facilities have also been considered briefly. In many cases the current imaging sensors and other perimeter systems may be adequate as available; in other cases they are being improved through research and development (R&D). Signature analysis for terrorist activities is a very difficult problem from a purely S&T point of view. However, gains may be possible by using some of the cross-cutting technologies. Examples include face recognition algorithms embedded in image sensor processors. Alternatively, more complex processing could be embedded in sensors that are designed to dramatically enhance performance by drawing on novel bioinspired architectures and on large databases of known terrorists. The committee did not include the acoustic, seismic, and radio frequency (RF) sensors used for perimeter defense in this chapter, but it discusses them in other chapters.

The remainder of this section briefly summarizes technologies for detecting nuclear weapons and radioisotopes, conventional explosives, chemical agents, and biological agents, along with the relevant cross-cutting technologies. The study was of short duration, and the committee does not claim completeness. The scope of the S&T covered by this study is so broad that a complete analysis would be a massive undertaking. The approach was to illustrate the types of technology employed and the various stages of development by using a number of examples.

SENSOR TECHNOLOGIES

Traditional Imaging Sensors

The committee first mentions the advanced, high-performance imaging systems that infuse all aspects of national security and defense and also have rel-

[2]For example, the S&T for the detection of nuclear weapons is principally a Department of Energy responsibility, with some responsibility assigned to the Defense Threat Reduction Agency. In another example the appropriations for funding the S&T related to the detection of chemical agents and biological agents have been assigned to the Joint Program Office for Chemical and Biological Defense. In this situation, the Army must be sufficiently involved and aware so that it can influence the S&T investments of other agencies and benefit from the results of those investments.

[3]The term cross-cutting technologies implies the merging of technologies that are being devel-

evance for HLS in I and W as well as in denial and survivability (Chapter 3). High-performance sensors, which image in a broad range of spectral bands, are a high priority for numerous theater and national missile defense platforms. The Department of Defense (DoD) in general and the Army specifically have broad programs in imaging sensors. Applications in addition to infrared (IR) imaging—such as techniques for sensing threats due to harmful chemical and biological agents—may be incorporated in different sensor suites. These system applications would require narrow spectral discrimination over broad spectral bands, low-light-level detection, increased sensitivity, and the ability to perform multifunctional imaging.

Detectors with responsivity in the IR atmospheric transmission band are desirable for the detection of terrestrial sources against a 300 K background. The main detectors currently available or in research include the following:

- HgCdTe imaging IR arrays,
- Uncooled bolometer arrays,
- GaAs quantum well arrays,
- GeSi internal photoemission detectors,
- GaSb intersubband and Type II detectors, and
- GaN detectors.

In addition to the thermal sensors described above, the Army has relied very heavily on night vision goggles as a primary imaging technology to support night operations. These goggles are used both for target acquisition and navigation, including pilotage. As an image intensification device, night vision goggles rely on the amplification of ambient light, such as starlight or moonlight. The Army is currently working on the fourth-generation image intensification device, with each generation of device becoming progressively smaller and more efficient.

The Army has broad programs in most of the above-listed detectors, particularly the first three, and there is ongoing research for improving their performance as well as for studying the causes and modes of degradation and failure. DARPA has several ongoing programs in lasers and nitride detectors for the ultraviolet and solar blind regions. In a situation where chemical or biological agents have been released into the atmosphere, this technology may be significant for standoff chemical and biological detection, as biological agents in particular have very specific signatures of absorption or emission in the ultraviolet portion of the spectrum.[4] Table 2-1 describes traditional imaging sensors.

---

oped independently and that are multidisciplinary in nature as well as perhaps being multiuse for a greater payoff.

[4]Detecting an aerosol cloud is much easier than characterizing what is in the cloud, but if the nature of the biological cloud is already known from other measurements, it should be possible to track the specific cloud and monitor its dispersion.

TABLE 2-1 Technologies for Perimeter Defense and Warning[a]

Technology	Characteristics	Availability[b] (R, N, F)	Priority for Army S&T[c]	Multiuse[d] (H, O, C)
HgCdTe imaging LWIR arrays[e]	Material of choice to fabricate high-performance detector arrays. Energy gap can be tailored in the range from 1.4 to 20 microns.	R	High	H, O, C
Uncooled bolometer arrays[e]	Utilizes temperature-dependent dielectric constants and operates at room temperature. $BaSiTiO_3$ (BST) is ferroelectric below a Curie temperature (Tc) of nearly 300 K. Current devices are optimized for response in the mid- and long-wave IR band, but in principle future bolometers can be made with a wide variety of responses using different absorptive coatings.	R-N	High	H, O, C
GaAs quantum well arrays[e]	A quantum well detector can be thought of as a type of extrinsic photoconductor in which the bound electrons reside inside the quantum wells instead of on dopant ions.	R-N	High	H, O, C

GeSi internal photoemission detectors	Makes use of the high internal photoemission of the GeSi alloy.	N-F	Low	H, O, C
GaSb intersubband and Type II detectors[f]	Potential for tunability and design	F	Low	H, O, C
GaN UV detectors for solar blind applications[g]	UV light selectively ionizes chemical agents. Ion detector determines concentration	F	High[h]	H, O, C

NOTE: LWIR, long-wave infrared; UV, ultraviolet.

[a]Impacts chemical and biological technologies.

[b]Availability: R, ready (TRL 8-9); N, near-term (TRL 4-7); F, far-term (TRL 1-3).

[c]Priority for Army S&T (investment): low, someone else has mission or technology is ready and available; medium, useful but of limited impact and some investment needed; high, very important, no one else working on it, considerable investment needed.

[d]Multiuse: H, Army homeland security; O, Objective Force; C, civilian (first responders and others).

[e]Westervelt et al. (1991).

[f]NRL (1998).

[g]DARPA (2002a,b).

[h]Does not include acoustic, seismic, and radio frequency sensors, which are additional perimeter defense technologies.

In the process of collecting material for this chapter, data on the performance of many different sensors were examined. As one would expect, the performance or utility of individual sensor technologies was dependent on the environment in which they were used. This led to some confusion in comparing performance among sensors. A consistent methodology would be helpful for reporting the performance of sensors in the environments in which they will actually be used. It makes little sense, for example, to present data on the sensitivity of a particular diagnostic methodology without also presenting the trade-off with specificity.[5] The NRC study *Making the Nation Safer* (NRC, 2002) calls for the following system-design approach:

- Establishment of standards for response time and field stability/durability, for example, for detection of WMD;
- Use of two-level sensor systems in which a low false-alarm-rate sensor with low specificity triggers a second sensor with a higher false-alarm rate but higher specificity;
- Use of multiple sensors and reasoning algorithms to obtain lower overall false-alarm probability, to predict contamination spread, and to provide guidance for recovery actions; and
- Use of networked sensors to provide wide-area protection of high-threat targets.

Conclusion 2-1. In conducting the survey it was often difficult to obtain authoritative and certified data on the real-world performance of many of the indicators and warning sensors in use or in development. This difficulty also applied to data on sensitivity and noise characteristics.

Recommendation 2-1. It is critically important that all sensors not only be well characterized at the point of purchase but also be regularly rechecked by competent technicians. Software used to integrate disparate sensors should be well documented and checked against standardized problems.

Chemical Agents

A number of different technologies are in use or in development for the detection of chemical agents. The agents are typically released by some means

[5]One elementary method of accomplishing this is through the use of receiver operating characteristic curves. These curves plot the true positive rate against the false positive rate under conditions appropriate for the test being made. Without such data it is difficult to draw meaningful conclusions from the measurements. The curves generally quantify how an increase in sensitivity is accompanied by a decrease in specificity. They are routinely used in evaluating sensor performance in a broad range of fields, from medical diagnostics to the design of radar systems.

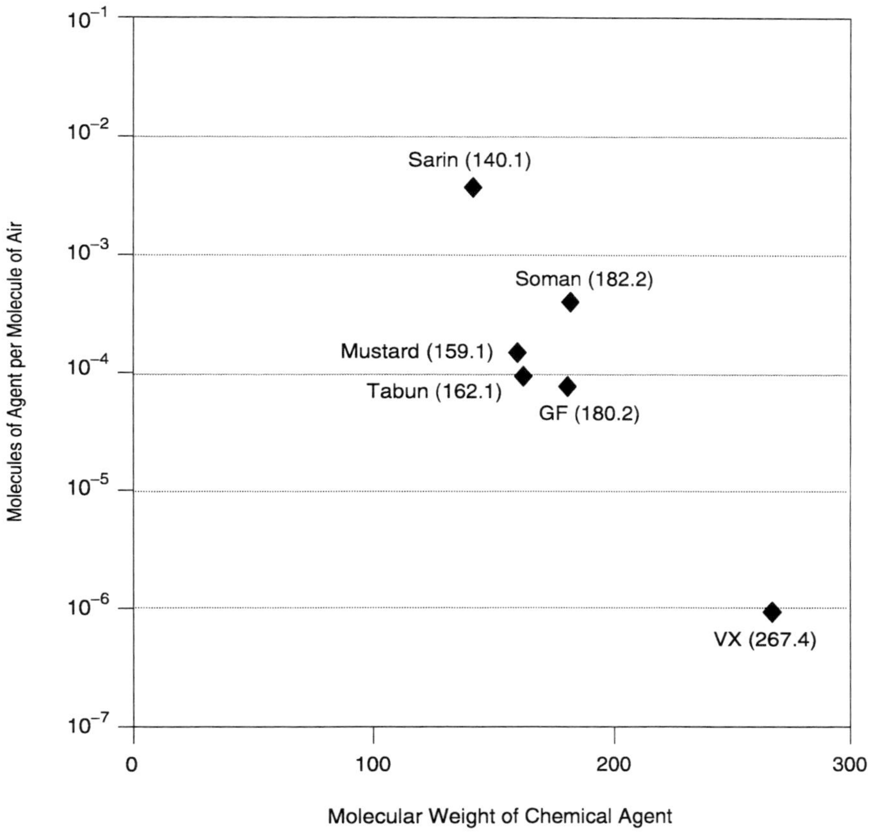

FIGURE 2-1 Vapor pressure concentrations for a number of chemical agents. SOURCE: Nerve agent data from Augerson (2000); mustard agent data from U.S. Army (undated).

into the atmosphere, where they form toxic clouds that are moved by atmospheric winds or by ventilation systems. The most desirable situation would detect these agents before they are released into the atmosphere. For weaponized agents this will be difficult. Figure 2-1 provides the vapor pressure concentrations for a number of chemical agents.

When compared with explosives, the chemical agents shown in Figure 2-1 are high-vapor-pressure substances. These concentrations will be easily detected with a number of technologies (however, VX will stress the state of the art for detection in realistic environments).

The acceptable exposure levels, however, are much lower than the vapor pressure levels. Figure 2-2 provides the atmospheric exposure limits (AEL) for a variety of chemical agents.

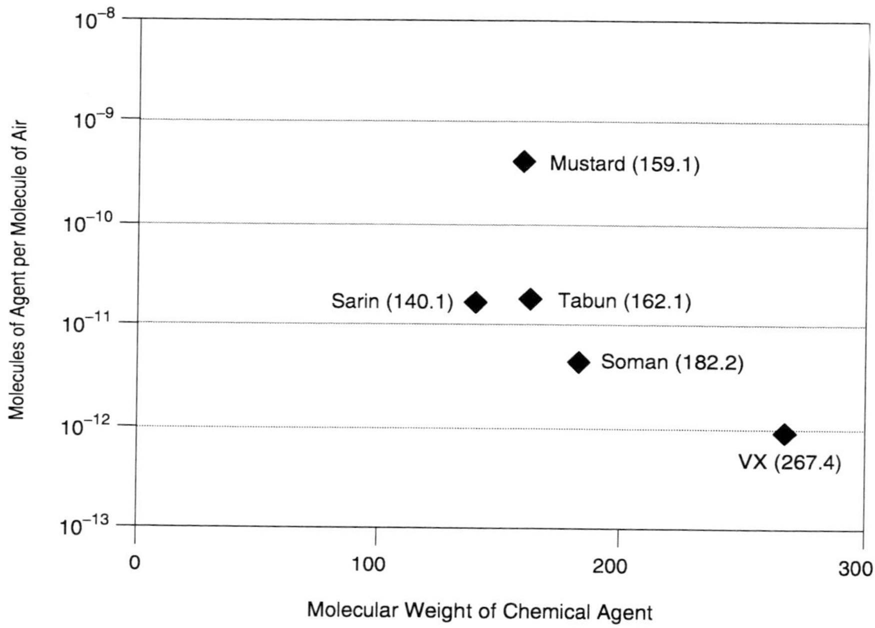

FIGURE 2-2 Atmospheric exposure limits for a variety of chemical agents. SOURCES: Nerve agent data take from Augerson (2000), CMS (undated); mustard agent data from U.S. Army (undated), CMS (undated).

These concentrations are more like those of the most-difficult-to-detect explosives, and one can expect similar problems with sensitivity and false alarms when operating in realistic, dirty environments. In clean environments where interfering substances can be kept to a minimum, the detection of trace amounts of chemical agents is more straightforward. Table 2-2 provides examples of means of chemical agent detection.

The use of industrial chemicals to cause harm should receive serious attention. If industrial chemicals are introduced into the atmosphere, they may be easier to detect than chemical warfare agents. At room temperature, chlorine, for example, has a vapor pressure an order of magnitude higher than air, while the vapor pressure of phosgene is about 50 percent higher than that of air; the vapor pressure of hydrogen cyanide is approximately the same as that of air, and the vapor pressure of methyl isocyanate is roughly half the vapor pressure of air. Also, because these chemicals are used routinely for industrial processing there is substantial experience in monitoring their presence at levels established by the Surgeon General as safe. The problem, of course, is that all of the monitoring is done in the industrial environments where these chemicals are expected to be present. Terrorists could employ these chemicals in locations where they would not be expected. This creates a sensor distribution problem.

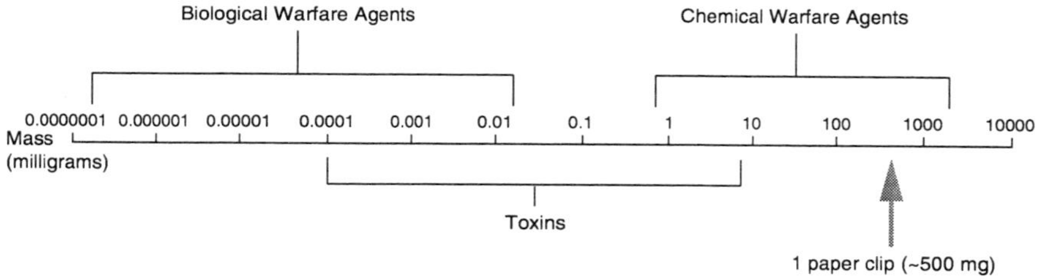

FIGURE 2-3 Comparative toxicity (amount needed to incapacitate) of biological agents, toxins, and chemical agents. SOURCE: NIJ (2001).

It should be noted that a number of these technologies also have relevance to the detection of conventional explosives and are therefore appropriate candidates for Army S&T investments for that purpose. Ion mobility spectrometry has broad application. The measurement presents a mass spectrum for fragments that are introduced into a drift chamber. Interpretation of this mass spectrum is where specific subject matter expertise comes into play. Interpretation of the spectrum for biological applications requires very different expertise than, say, interpretation of the spectrum collected in an explosives detection test. This technology could certainly be considered as cross-cutting in much the same sense that quantum dots technology is cross-cutting in its applications.

Biological Agents

The point detection of biological agents is qualitatively different from that of chemical agents. This is seen in Figure 2-3, which compares the amount of biological agent needed to incapacitate an individual with the amounts of chemical agent and toxin needed to incapacitate. Many orders of magnitude less biological agent is required.

Most devices for the physical detection of biological agents require that the agent be in the environment. A typical biodetection system involves a queuing, detection, discrimination, and identification sequence. This sequence requires that samples be purified and concentrated so that other species that could potentially interfere with detection of the target agent are reduced to a minimal level. Some of the technologies that are utilized or are under investigation for implementing this sequence are listed in Table 2-3.

There are many promising opportunities for investing S&T funding in support of biological agent detectors. Responsibility for this area, however, has been assigned to the Joint Project Office for Chemical Biological Defense. This limits the investment of Army S&T funding in this important area. There are, however, some undertakings also relevant to explosives detection and many other undertakings relevant to the Objective Force that are within the purview of the DASA (R&T) and that can help advance biodetection S&T. Some of these will be mentioned the next section.

TABLE 2-2 Technologies for Chemical Agent Detection

Technology	Characteristics	Availability[a] (R, N, F)	Priority for Army S&T[b]	Multiuse[c] (H, O, C)
Enzymatic paper	Detects nerve gas at ~ppb, HC at ~10 ppm, and mustard gas at ~ppb. Inexpensive, prone to false positives.	R	Low	H, O
Ion mobility spectroscopy	Detects nerve gas at ~6 ppb and mustard gas at ~10 ppb. 1-2 minutes. Erroneous detection from interference, e.g., smoke.	R-N	Medium[d]	H, O, C
Photo acoustic IR spectroscopy	Highly selective. Sensitive to external vibration.	R-N	Medium[d]	H, O, C
Differential absorption light detection	Tracks identified clouds. Sensitive to environmental noise.	R-N	Low	H, O, C

Passive IR detection	Direct measurement of IR emission or absorption from chemical agent cloud.	R-N	Low	H, O, C
Photo ionization	UV light selectively ionizes chemical agents. Ion detector determines concentration.	R-N	Medium[d]	H, O, C
Flame photometry	Flame color determines concentration of sulfur and phosphorous. Highly sensitive. Prone to false positives.	R-N	Low	H, O, C
Gas chromatography	Vapor separation through a column improves flame photometry.	R-N	Medium[d]	H, O, C
Surface acoustic wave	Surface absorption of chemical agents changes resonance frequency. Measures many chemical agents simultaneously.	R-N	Medium	H, O, C

NOTE: ppb, parts per billion; ppm, parts per million; UV, ultraviolet.

[a]Availability: R, ready (TRL 8-9); N, near-term (TRL 4-7); F, far-term (TRL 1-3).

[b]Priority for Army S&T (investment): low, someone else has mission or technology is ready and available; medium, useful but of limited impact and some investment needed; high, very important, no one else working on it, considerable investment needed.

[c]Multiuse: H, Army homeland security; O, Objective Force; C, civilian (first responders and others).

[d]Impacts chemical and biological technologies.

SOURCE: Davis and Kelen (2001).

TABLE 2-3 Technologies for Biological Agent Detection

Technology	Characteristics	Availability[a] (R, N, F)	Priority for Army S&T[b]	Multiuse[c] (H, O, C)
Aerodynamic particle sizing[d, e]	Measures relative number of particles in a given size range. Nonspecific, empirical.	R	Low	H, O, C
Fluorescence particles sizing[d, e]	Measures relative number of particles in a given size range and discriminates between nonbiological entities and biological entities.	R	Low	H, O, C
Flow cytometry[d, e]	Measures physical and chemical characteristics of cells.	R	Low	H, O, C
Pyrolysis[e]	Uses controlled rapid heating to decompose complex organic molecules into fragments that may have distinct chemical signatures.	R	Low	H, O, C
Mass spectrometry[d]	Determines structure and molecular weight of biomolecule fragments.	R-N	Low	H, O, C
Gas chromatography[e]	Separates components in gaseous mixture.	R-N	Medium[f]	H, O, C
Ion mobility spectrometry[e]	Measures ion drift times through buffer gas in drift tube.	R-N	Medium[f]	H, O, C
Flame photometry[e]	Measures phosphorus emission lines from gas-phase biomolecule fragments.	R	Low	H, O, C

Immunoassay[d, e]	Detects and measures specific binding of antigens with their corresponding antibodies or gene abundance.	R-N	Medium[f]	H, O, C
DNA microarrays	Monitors thousands of genes simultaneously.	F	High[f]	H, O, C
Nucleic acid amplification[d, e]	Uses unique DNA structure of biological organism to identify pathogens and BW agents.	R-F	Low	H, O, C
Combinatorial peptides	Uses massive libraries for screening.	F	High	H, O, C
Capillary electrophoresis[e]	Allows rapid separation of ions and subsequent detection of separated species.	R	Low	H, O, C
Ion channel switch[e]	Exploits selective movement of ions across biological membranes.	F	Low	H, O, C
Cell-based[e]	Exploits electrical activity of cells to detect broad range of agents.	F	Low	H, O, C
Raman scattering[e]	Matches observed Raman spectra against library of predetermined signatures.	N-F	High[f]	H, O, C

NOTE: BW, biological warfare; PCR, polymerase chain reaction.

[a]Availability: R, ready (TRL 8-9); N, near-term (TRL 4-7); F, far-term (TRL 1-3).

[b]Priority for Army S&T (investment): low, someone else has mission or technology is ready and available; medium, useful but of limited impact and some investment needed; high, very important, no one else working on it, considerable investment needed.

[c]Multiuse: H, Army homeland security; O, Objective Force; C, civilian (first responders and others).

[d]NIJ (2001).

[e]NATIBO (2001).

[f]Impacts chemical and biological technologies.

Nuclear Materials

In the case of nuclear weapons, the primary fissionable isotopes of interest are uranium-235, plutonium-239, and uranium-233. Consider, for example, a plutonium weapon: the signatures from spontaneous decay processes will be gamma rays and neutrons, which are detectable at a distance. Assuming scattering but no neutron capture between the weapon and the detector, the weapon neutron flux from spontaneous fission will equal the background neutron flux at about 15 m from the weapon.[6] If one wishes to detect at a longer distance the spontaneous neutron output from the plutonium weapon, one must deploy detectors capable of detecting excess thermal neutrons at levels below the background flux level. A similar situation exists for gamma radiation from plutonium. However, U-235 is more difficult to detect since it has a low spontaneous fission rate and therefore does not provide a strong neutron signal. It does have a low-energy gamma-ray emission spectrum with well-understood structure. This gamma-radiation emission spectrum is used for detecting and identifying U-235-based weapons. The detection range is quite limited due to the low gamma-ray energy and the natural background of gamma radiation. This problem has been studied for many years, and a variety of technologies have emerged, some of which are quite well understood and others of which are relatively new. Uranium-233 is of marginal interest because there is so little of it in the world. Table 2-4 describes technologies that are in development or currently used for the detection of neutrons and gamma rays in the nuclear weapons context.

The detection ranges for these technologies are relatively short, and they are best deployed in a choke point or a portal situation. This area is extremely important, but responsibility for conducting the appropriate S&T resides with the Department of Energy and the Defense Threat Reduction Agency (DTRA). As a result, the appropriateness of expending Army S&T funds was rated low. However, the impact of Army S&T investments in areas such as perimeter defense, miniaturized sensor technology, networked sensors, and data fusion could have a great influence on detection of nuclear materials.

The principal objective of a radiological dispersion weapon ("dirty bomb") is to spread radioactive material by detonating a conventional explosive in proximity to the radioactive material or by spreading radioactive material as an aerosol. Any radioactive material could be used for this purpose. It is expected that the main source of such weapons would be materials used for hospital radiation therapy (such as iodine-125, cobalt-60, or cesium-137), radio pharmaceuticals (such as iodine-131, iodine-123, technetium-99, and xenon-134), or nuclear power

---

[6]It takes 1 to 8 kg of Pu-239 to make a plutonium weapon. Weapons-grade plutonium will contain a few percent of Pu-240, which has a high spontaneous fission rate, resulting in the emission of about 1 million neutrons per second (per kg of Pu-240). Therefore a nuclear weapon containing 5 kg of weapons-grade plutonium will emit about 3×10^5 neutrons/sec. The natural background of thermal neutrons is about 10^{-2}neutrons/sec-cm^2 (NSSS, 2000).

plant spent fuel rods, which contain fission products. These materials are principally gamma-ray emitters, and the detection would involve gamma-ray detection technologies such as those listed in Table 2-4.

All of the nuclear materials detectors mentioned above have relatively short detection ranges and are best suited for choke point or portal geometries or where there is good intelligence on where the material is located. The same will be true for the detectors of the other substances to be discussed in this section. It seems unlikely that the nation can afford to create, equip, and staff enough portals to make negligible the probability of dangerous materials entering the country.

Conventional Explosives

The majority of terrorist attacks against U.S. forces, facilities, and citizens have involved the use of conventional explosives. The detection and tracking of these explosives are therefore of great importance for HLS and are highly applicable to the Objective Force. Conventional explosive detection technologies generally fall into two categories: vapor-phase detection and bulk detection. Figure 2-4 provides some insight into the vapor pressures of the better-known explosives.

As can be seen from Figure 2-4, the vapor pressure of explosives varies over a wide range, with the older explosives having vapor pressures measured in parts per million (ppm) relative to atmospheric pressure and the more modern explosives having vapor pressures in the range of parts per trillion (ppt).

A number of technologies are under development to examine the feasibility of detecting the vapor phase of explosives. Table 2-5 describes examples of devices that work by such detection.

It should be quite clear from Figure 2-4 and Table 2-5 that for modern explosives, vapor-phase detection of explosives will be limited to detectors in close proximity to the explosives or will require very substantial concentration of the explosive vapors at a distance from the explosive. At these very low detection levels, interfering species will clearly be a big issue.

Army weapons and explosives in transit or in storage can be an attractive target for theft or diversion by terrorists. Surface-to-air missiles, antivehicle weapons, mines, and bulk explosives are particularly well suited for terror attacks. On a broader scale, it would be in the interest of the United States if international protocols were established that call for the insertion of detection markers[7] and identification taggants[8] into all legitimately manufactured explosives worldwide to assist both detection and forensic analysis. This was discussed in the NRC report *Containing the Threat from Illegal Bombings* (1998).

---

[7]Detection markers are materials added to explosive that can be sensed before a blast by an instrument designed for that purpose (NRC, 1998).

[8]Identification taggants are additives designed to survive an explosive blast, to be recoverable at the site of a bombing, and to provide pertinent information (NRC, 1998).

TABLE 2-4 Technologies for the Detection of Neutrons and Gamma Rays in the Nuclear Weapons Context

Functionality	Technology	Characteristics	Availability[a] (R, N, F)	Priority for Army S&T[b]	Multiuse[c] (H, O, C)
Passive gamma-ray detection	Sodium iodide crystals and PMT[d]	Robust, high false-alarm rate at high sensitivity, poor spectral resolution.	R	Low	H, O, C
	Germanium crystals[d]	Sophisticated instrument, needs refrigeration, very high resolution (does not make mistakes).	R	Low	H, O, C
	Mercuric iodide[e]	Intermediate spectral resolution, solid state, robust, relatively expensive.	R-N	Low	H, O, C
	CdZnTe[f]	Intermediate spectral resolution, solid state, relatively expensive.	R-F	Low	H, O, C
Passive neutron detection	Silicon strip[f]	Solid state, good energy resolution, 30° C.	F	Low	H, O, C
	Scintillating glass fibers[h]	Robust, good energy resolution, room temperature.	N-F	Low	H, O, C
	Scintillating glass fibers[h]	Robust, solid state, fabricated to desired geometry.	R-N	Low	H, O, C
	CMOS/SOI[d]	Relatively inexpensive, small, no database on utility in real world.	N-F	Low	H, O

Nonspecific particle counter	Geiger counters, simple ionization chamber[d]	Not specific, low sensitivity, relatively inexpensive.	R	Low	H, O, C
Neutron spectroscopy	Pulsed neutron source[i]	Sophisticated instrument, radiation hazard.	R-N	Low	H, O
Active gamma-ray scanner	Pulse power or radioactive gamma source[j]	Robust technology, radiation hazard, good spatial resolution and imaging.	R	Low	H, O, C

NOTE: PMT, photomultiplier tube; CMOS, complementary metal oxide semiconductor; SOI, silicon on insulator.

[a]Availability: R, ready (TRL 8-9); N, near-term (TRL 4-7); F, far-term (TRL 1-3).

[b]Priority for Army S&T (investment): low, someone else has mission or technology is ready and available; medium, useful but of limited impact and some investment needed; high, very important, no one else working on it, considerable investment needed.

[c]Multiuse: H, Army homeland security; O, Objective Force; C, civilian (first responders and others).

[d]LBNL (2001); personal communications between R. Whitlock and R. August, U.S. Naval Research Laboratory, and Tim Coffey, Committee on Army Science and Technology for Homeland Defense, November 2001.

[e]Contech (2000).

[f]LBNL (2002); personal communications between J. Kurfess, U.S. Naval Research Laboratory, and Tim Coffey, Committee on Army Science and Technology for Homeland Security, April 2002.

[g]LBNL (2002); personal communications between J. Kurfess, U.S. Naval Research Laboratory, and Tim Coffey, Committee on Army Science and Technology for Homeland Security, April 2002.

[h]NSSS (2000).

[i]LBNL (2001) and Seymour et al. (1999).

[j]SAIC (2002).

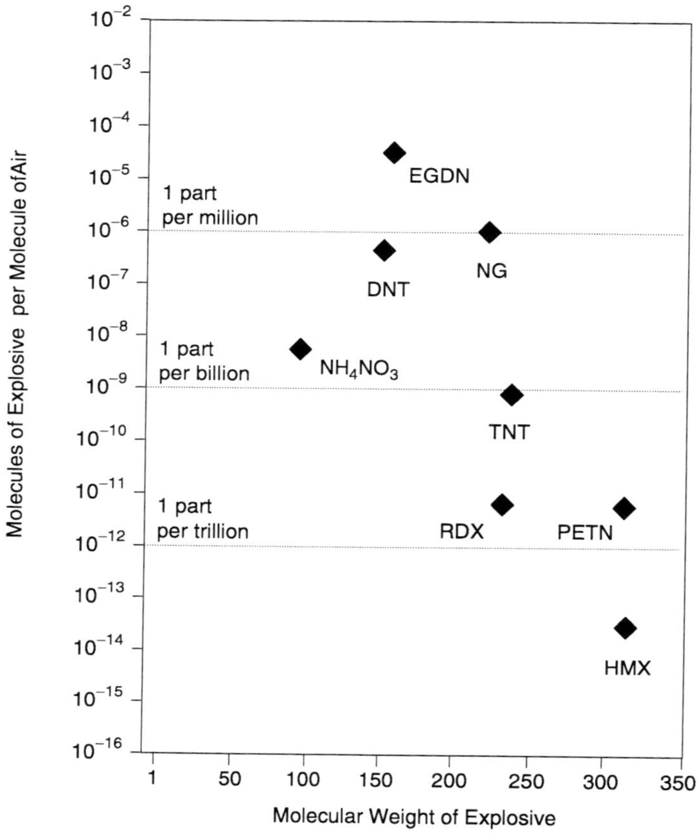

FIGURE 2-4 Vapor pressure associated with the better-known explosives. SOURCE: Adapted from NIJ (1999).

Modern explosives manufactured to include higher-vapor-pressure taggants will have longer detection ranges. The inclusion of such markers will, of course, make military explosives somewhat easier to detect, which may have implications for operational security (OPSEC). However, since the black market in military explosives is of concern to HLS, the OPSEC implications of markers may represent an acceptable trade-off.

Conclusion 2-2. Technologies should be pursued that (1) deny theft or diversion by maintaining real-time inventory control, then tracking if control is lost or (2) reduce the utility of such equipment to terrorists. Incorporation of detection markers and identification taggants into all legitimately manu-

TABLE 2-5 Technologies for Vapor-Phase Explosive Detectors

Technology	Characteristics	Availability[a] (R, N, F)	Priority for Army S&T[b]	Multiuse[c] (H, O, C)
Ion mobility spectrometer[d,e]	Detects at parts per billion level. Must be close to explosive or chemical. Noise limits become a problem at low signal levels. Fundamental problem in selectivity and resolution. Shows promise for increased detection in low concentrations.	R-N	Medium	H, O, C
Chemical resistors[e,f]	Detects at parts per billion level. Must be close to explosive or chemical, needs improved SNR.	N	High	H, O, C
Fluorescent polymers[d]	Detects at parts per trillion level (in principle). Must be close to explosive or chemical, needs improved SNR. Demonstrated at parts per billion in reliable system.	R-N	High	H, O, C
Gas chromatography + SAW[d,g]	Detects at parts per billion level. Must be close to explosive or chemical, must be able to desorb the explosive vapors for system to be useful.	R-N	Medium	H, O, C
Surface-enhanced Raman spectroscopy[d]	Detects at parts per billion. Portable, must be close to explosive.	N-F	High	H, O, C
Immunoassay (biosensors)[d]	Detects parts per billion. Must be close to explosive. Potential for increased sensitivity.	N-F	High	H, O, C

NOTE: SNR, signal-to-noise ratio; SAW, surface acoustic wave.

[a]Availability: R, ready (TRL 8-9); N, near-term (TRL 4-7); F, far-term (TRL 1-3).

[b]Priority for Army S&T (investment): low, someone else has mission or technology is ready and available; medium, useful but of limited impact and some investment needed; high, very important, no one else working on it, considerable investment needed.

[c]Multiuse: H, Army homeland security; O, Objective Force; C, civilian (first responders and others)

[d]Ward et al. (2001).

[e]Lewis et al. (1997).

[f]Bruschini and Gros (1997).

[g]U.S. Navy (2002).

factured low-vapor-pressure explosives will assist in both detection and forensic analysis.

Recommendation 2-2. An international convention requiring the incorporation of detection markers and identification taggants should be sought.

Bulk-phase detection of explosives generally involves some form of interrogation of the explosive. All of the systems require close proximity to the material being interrogated. Table 2-6 describes examples of bulk explosive detection.

Conclusion 2-3. The physical detection of dangerous packaged materials (nuclear weapons, radiological weapons, chemical weapons, biological weapons, and explosive weapons) is an extremely difficult and stressing task, even when the materials are forced through choke points.

CROSS-CUTTING TECHNOLOGIES

It is quite clear that the great majority of technologies for the physical detection of nuclear weapons, radiological weapons, conventional explosives, chemical agents, and biological agents require close proximity to the weapon. Detection of chemical or biological aerosol clouds at a distance is possible. However at that point, the attack is already under way. Similarly, the use of health and medical surveillance, while very desirable, is a post-attack undertaking. The most desirable indication and warning would signal the presence of dangerous material before an attack has begun. While efforts should continue to improve pre-event detection ranges for individual sensors, it is clear that the laws of physics, chemistry, and biology will impose severe limits on these ranges. This would seem to leave two options for the physical detection of dangerous materials:

- One option is to force all material to move through choke points or portals. This will bring the detectors and the dangerous materials into proximity, thereby easing the burden on detector technology.
- The second option would involve distributing large numbers of detectors, making it difficult to avoid detection by avoiding choke points and portal systems. This second option would require inexpensive detectors that can be widely proliferated. It would also require sophisticated networking of the detectors and the development of systems to intelligently interpret the data provided by them.

The distributed network would involve fixed sensors and mobile sensors deployed on various platforms, including autonomous unmanned air, space, ground, and underwater vehicles. This option opens up substantial opportunities for the investment of Army S&T resources because the S&T involved is applicable to the Army for more than just nuclear weapons detection or chemical and biological agent detection. For example, the intelligent networking of sensors involves S&T that cuts across many applications of interest to the Objective Force, including perimeter defense, tracking, identification, and targeting. Similarly, the S&T needed to develop inexpensive small sensors for wide proliferation would involve studies that are much broader than those specific to HLS. Indeed, the most significant advances in detection technologies may come from the innovative combination of very disparate technologies into compact integrated sensor suites. The S&T for the required autonomous unmanned sensor platforms is of great interest to the Objective Force and will have an important impact. Learning how to do all of this will be of very broad interest to the Army.

In addition to existing or anticipated ideas for detection, the committee thought it worthwhile to highlight more speculative means for detection in Box 2-1.

There are many examples where cross-cutting technologies have had an impact well beyond that initially envisioned. Consider the case of fiber-optic sensors. These were originally developed by the DoD to provide for the sensitive detection of acoustic, magnetic, and strain signatures. In one variation, these detectors utilize evanescent field excitation, whereby a portion of the light traveling in the fiber core penetrates the surrounding medium with the power of the evanescent field decaying exponentially from the fiber core. Through a clever combination of surface chemistry, biological or chemical receptors can be bound to the surface of the cladding. By introducing a fluorophore into this arrangement and monitoring the change in fluorescence that occurs when specific binding takes place at the surface of the fiber it was possible to create a fiber-optic detector for certain chemicals and biological entities. This is an example where S&T developed by DoD for purposes having nothing to do with chemical or biological detection has made an important contribution to the detection of biological agents.

As another example, consider the S&T that has been supported by DoD in semiconductor quantum-dot nanocrystals. These quantum dots have been shown to have emission spectra that may be tuned by changing the quantum-dot radius. For example, quantum dots may be fabricated so that a 2-nanometer particle glows bright green while a larger 5-nanometer particle glows red in the presence of white light. These developments originally had nothing to do with the detection of chemical or biological agents, but the dual-use potential was found through clever chemistry. The utility of this approach is limited by the efficiency of the immunoassay or the DNA identification technique. It remains to be seen whether or not a viable detection system can be developed for quantum dots.

It should be clear from the above discussion that the cross-cutting technologies could have a broad impact and should be of very great interest to the DASA (R&T). Some examples of relevant cross-cutting technologies are shown in Table 2-7.

TABLE 2-6 Technologies for Bulk Explosive Detection

Technology	Characteristics	Availability[a] (R, N, F)	Priority for Army S&T[b]	Multiuse[c] (H, O, C)
Transmission x-ray[d,e]	Portal system provides 2-D images.	R	Low	H, O, C
Transmission gamma ray[e]	Portal system provides 2-D images.	R	Low	H, O, C
Backscatter x-ray[e]	Finds low-atomic-number elements (C, H, O, N). Requires close proximity and sophisticated interpretation.	R-N	Low	H, O, C
X-ray and gamma-ray tomography[e]	Portal system provides 3-D images.	R	Low	H, O, C
Thermal neutron analysis (TNA)[d,e,f]	Portal system: capture of thermal neutron by nitrogen gives 10.8 MeV gamma ray.	R-N	Low	H, O, C
Fast neutron analysis (FNA)[e]	Portal system: stimulates gamma radiation from elements being irradiated.	R-N	Low	H, O, C

Pulsed fast neutron analysis (PFNA)[e,f]	Portal system: stimulates gamma radiation from elements being irradiated.	R-N	Low	H, O, C
Nuclear magnetic resonance (NMR)[e f]	All samples must be passed through magnetic coils. Chemical interpretation of NMR transitions can determine composition.	R	Low	H, O, C
Nuclear quadrupole magnetic resonance (NQR)[e,f]	Low SNR, must be close to explosive, does not require magnets. Produces RF signals characteristic of particular explosives.	R-N	High	H, O, C
Millimeter-wave radiometry[f,g]	Potential to provide radiometric images of objects (e.g., explosives) under clothing.	N	High	H, O, C

NOTE: 2-D, two-dimensional; 3-D, three-dimensional; MeV, mega-electron-volt; RF, radio frequency.

[a]Availability: R, ready (TRL 8-9); N, near-term (TRL 4-7); F, far-term (TRL 1-3).

[b]Priority for Army S&T (investment): low, someone else has mission or technology is ready and available; medium, useful but of limited impact and some investment needed; high, very important, no one else working on it, considerable investment needed.

[c]Multiuse: H, Army homeland security; O, Objective Force; C, civilian (first responders and others).

[d]Ward et al. (2001).

[e]U.S. Navy (2002).

[f]Bruschini and Gros (1997).

[g]NRC (1996).

TABLE 2-7 Examples of Cross-Cutting Technologies

Application	Technology	Characteristics	Availability[a] (R, N, F)	Priority for Army S&T[b]	Multiuse[c] (H, O, C)
Detection and tracking	Sensor networking (see Box 2-1)	Gathers data from a wide variety of spatially distributed sensors.	N-F	High	H, O, C
	Sensor fusion	Intelligently combines, correlates, and interprets data from distributed sensors.	N-F	High	H, O, C
	Anomaly detection	Examines data from networked sensors to discover patterns, unusual behavior, etc.	N-F	High	H, O, C
	Surveillance platforms (UAVs, UGVs, UUVs)	Small autonomous vehicles for carrying sensor payloads as part of distributed sensor network.	R-F	High	H, O, C
Perimeter surveillance	IR, RF, acoustic, seismic, etc. techniques	Monitors for intrusion into predetermined spaces (encampments, facilities, borders, etc.).	R-N	High	H, O, C

I and W capability in miniaturized systems	MEMS	Methods for integration of many technologies into microsensors using electronic fabrication technologies.	R-F	High	H, O, C
	Active-passive sensor suites	Suites of lasers and detectors that can query and image as well as perform spectroscopic measurements.	N-F	High	H, O, C
	Nanofabrication techniques	Fabrication of sensing systems at the atomic level.	F	High	H, O, C

NOTE: UAV, unmanned air vehicle; UGV, unmanned ground vehicle; UUV, unmanned underwater vehicle; IR, infrared; RF, radio frequency; MEMS, microelectromechanical systems.

[a]Availability: R, ready (TRL 8-9); N, near-term (TRL 4-7); F, far-term (TRL 1-3).

[b]Priority for Army S&T (investment): low, someone else has mission or technology is ready and available; medium, useful but of limited impact and some investment needed; high, very important, no one else working on it, considerable investment needed.

[c]Multiuse: H, Army homeland security; O, Objective Force; C, civilian (first responders and others).

BOX 2-1
Speculation on Means of Detection Using the Existing Telecommunications Infrastructure

The committee notes in Table 2-4 that glass fibers subjected to gamma radiation near background levels scintillate. Although the scintillation is weak, it is detectable and the effect is used to detect gamma radiation. The telecommunications industry has introduced a good deal of glass fiber into the country's infrastructure. Those fibers that are above ground undoubtedly exhibit some level of optical noise due to the gamma radiation background. This background radiation level will increase if a gamma radiation source approaches the fiber. If this were detectable, then the telecommunication optical fiber infrastructure might itself serve as a distributed network of gamma-radiation detectors.

As another example, consider the fact that the natural background of thermal neutrons has been shown to cause single-event upsets in microelectronics. The thermal neutrons interact with the boron-10 fraction of boron dopants, producing alpha particles. The energy deposited by the alpha particles causes the upsets. Perhaps this effect could be exploited to produce a highly distributed thermal neutron detection system by incorporating a special boron-doped chip in cell phones. When a phone "shakes hands" with a cell tower, it could pass a neutron anomaly message and its GPS coordinates, if equipped to do so. If something like this were feasible it would result in a worldwide distributed network of thermal neutron detectors.

SUMMARY

A new approach is required for the indication and warning stage for chemical, biological, radiological, nuclear, or high explosive weapons. There are many opportunities for the Army S&T program to help in defining that new approach. The new approach might involve the proliferation of small but competent sensor systems into some sort of intelligent network. Exploitation of the nation's existing infrastructure should be examined. Such an undertaking would require expanding the community currently working on indications and warning. The collective skills of this community might enable a new class of detector system that makes it difficult to position terrorist weapons so that they are a threat to U.S. forces or to the general population. This distributed sensors approach offers many important opportunities for investigation by the Army S&T program.

The Army's role in funding S&T for detectors of CBRNE weapons is very limited. There are, however, numerous opportunities for synergy among legitimate Army S&T investments and the investments of others in detector technologies. This is especially true of cross-cutting technologies.

Many important contributions to I and W sensor capability are likely to come from developments in fields not traditionally associated with CBRNE weapons or the detection thereof. The stovepipe communities,[9] funding agencies, and funding mechanisms that have been set up in CBRNE weapons areas, while very effective in cases where it is known how to solve a problem, can be counterproductive in this situation.

The interrelationships needed among the sensor networks and for the broader intelligence collection activity are difficult to establish, for technical, cultural, and legal reasons. Nevertheless, the committee envisioned a situation where the relevant sensor networks would be queued as a result of intelligence findings, with the intelligence community tasked to undertake focused collection efforts if the sensor networks picked up unusual activity. There are serious scientific and technical questions here even if the cultural and legal issues can be resolved. For example, the ability to quickly and reliably search massive databases for anomalous activity would be critical for the implementation of this recommendation. It may be necessary to create a research organization to resolve this problem, and it is unlikely that any one institution can take this on. A consortium approach might work, but it would be confronted by serious if not insurmountable security classification problems.

Conclusion 2-4. A purely technical solution to the indications and warning problem based upon sensors, even networked sensors, is unlikely. Establishing the proper interrelationships among the sensor networks and the broader intelligence collection activity will be crucial for properly queuing the sensor network.

Recommendation 2-4a. The Army should ensure from the outset that the necessary interrelationships among the sensor networks and the broader intelligence collection activity are established and maintained as a coherent undertaking.

Recommendation 2-4b. Army science and technology should aggressively seek out and invest in those cross-cutting sciences and technologies that will benefit both the Objective Force and the homeland security requirement to detect weapons of mass destruction.

---

[9]A "stovepipe" community is a relatively closed community where certain franchises have been granted. These communities tend to be insular in terms of their involvement with larger communities, but they can be multidisciplinary. It is often very difficult for an outsider to break into these communities. They can be very effective when one knows how to solve a particular problem and it is simply a matter of assembling a team to get it done. They are less effective where solutions are not obvious and where truly new ideas are required. In the case of homeland security new ideas are clearly needed, and the government should be seeking the broadest possible involvement until a solution is at hand.

REFERENCES

Augerson, W.S. 2000. A Review of the Scientific Literature as it Pertains to Gulf War Illnesses, Volume 5: Chemical and Biological Warfare Agents. Available online at <http://www.rand.org/publications/MR/MR1018.5/MR1018.5.pdf/>. Accessed on October 10, 2002.

Bruschini, C., and B. Gros. 1997. A Survey of Current Sensor Technology Research for the Detection of Landmines. Available online at <http://diwww.epfl.ch/lami/detec/susdemsurvey.html>. Accessed on September 24, 2002.

CMS (CMS Field Systems). Undated. Chemical Warfare Agent Air Monitoring Systems. Available online at <http://www.nbcindustrygroup.com/cms.htm>. Accessed on October 10, 2002.

Contech (Constellation Technology). 2000. Mercuric Iodide Detectors. Available online at <http://www.contech.com/Mercuric_Iodide_Catalog..htm>. Accessed on September 24, 2002.

DARPA (Defense Advanced Research Projects Agency). 2002a. Semiconductor Ultraviolet Optical Sources (SUVOS). Available online at <http://www.darpa.mil/mto/suvos/index.html>. Accessed on October 2, 2002.

DARPA. 2002b. Solar Blind Detectors. Available online at <http://www.darpa.mil/MTO/SBD/index.html>. Accessed on October 2, 2002.

Davis, G., and G. Kelen. 2001. CBRNE–Chemical Detection Equipment, October 15. Available online at <http://www.emedicine.com/emerg/topic924.htm>. Accessed on September 23, 2002.

LBNL (Lawrence Berkeley National Laboratory). 2001. Nuclear Science—A Guide to the Nuclear Science Wall Chart: Tools of Nuclear Science. Available online at <http://www.lbl.gov/abc/wallchart/teachersguide/pdf/Ch12-toolsofNuclear%20Sci%20doc.pdf>. Accessed on September 24, 2002.

LBNL. 2002. Radiation Detectors Capabilities. Available online at <http://engineering.lbl.gov/cap/capdetail.asp?CapCode=RadDet>. Accessed on September 24, 2002.

Lewis, N.S., M.C. Lonergan, E.J. Severin, B.J. Doleman, and R.H. Grubbs. 1997. Array-based vapor sensing using chemically sensitive carbon black-polymer resistors. Pp. 660-670 in Detection and Remediation Technologies for Mines and Minelike Targets II, Proceedings of SPIE, vol. 3079, A.C. Dubey and R.L. Barnard, eds. Bellingham, Wash.: The International Society for Optical Engineering.

NATIBO (North American Technology and Industrial Base Organization). 2001. Biological Detection System Technologies Technology and Industrial Base Study, February. Available online at <http://www.dtic.mil/natibo/>. Accessed on September 23, 2002.

NIJ (National Institute of Justice). 1999. Guide for the Selection of Commercial Explosives Detection Systems for Law Enforcement Applications, NIJ Guide 100-99, December. Washington, D.C.: National Institute of Justice.

NIJ. 2001. An Introduction to Biological Agent Detection Equipment for Emergency First Responders, NIJ Guide 101–00, December. Available online at <http://www.ncjrs.org/pdffiles1/nij/190747.pdf>. Accessed on September 23, 2002.

NRC (National Research Council). 1996. Airline Passenger Security Screening: New Technologies and Implementation Issues. Washington, D.C.: National Academy Press.

NRC. 1998. Containing the Threat from Illegal Bombings: An Integrated National Strategy for Marking, Tagging, Rendering Inert, and Licensing Explosives and Their Precursors. Washington, D.C.: National Academy Press.

NRC. 2002. Making the Nation Safer: The Role of Science and Technology in Countering Terrorism. Washington, D.C.: National Academies Press.

NRL (Naval Research Laboratory). 1998. InAs/InGaSb Infrared Lasers and Detectors. Available online at <http://sinh.nrl.navy.mil/code6870/reprints/las&det.PDF>. Accessed on October 2, 2002.

NSSS (Nuclear Safeguards and Security Systems LLC). 2000. Why Neutrons. Available online at <http://www.nucsafe.com/puma/why_neutrons.htm>. Accessed on September 24, 2002.

SAIC (Science Applications International Corporation). 2002. Portal VACIS™ Technical Specifications. Available online at <http://www.saic.com/products/security/portal-vacis/portal-vacis-tech.html>. Accessed on September 24, 2002.

Seymour, R.S., R.A. Craig, M. Bliss, B. Richardson, C.D. Hull, and D.S. Barnett. 1999. Performance of a neutron-sensitive scintillating glass fiber panel for portal, freight, and vehicle monitoring. Pp. 148-155 in Nuclear Waste Instrumentation Engineering, Proceedings of SPIE, vol. 3536, D.E. Robertson, ed. Bellingham, Wash.: The International Society for Optical Engineering.

U.S. Army. Undated. U.S. Army Center for Health Promotion and Preventive Medicine Detailed Facts About Sulfur Mustard Agents H and HD. Available online at <http://chppm-www.apgea.army.mil/dts/docs/dethhd.pdf>. Accessed on October 10, 2002.

U.S. Navy. 2002. Department of the Navy Explosive Detection Equipment Program–Explosives. Available online at <http://explosivedetection.nfesc.navy.mil/explosives.htm>. Accessed on September 24, 2002.

Ward, K.B., A. Ervin, J.R. Deschamps, and A.W. Kusterbeck. 2001. Force protection: Explosives detection experts workshop, NRL/MR-MM/6900—01-8564, CDROM. Arlington, Va.: Office of Naval Research.

Westervelt, R., J. Sullivan, and N. Lewis. 1991. Imaging Infra-red Detectors. JASON report number JSR-91-600. McLean, Va.: Mitre Corporation.

3

Denial and Survivability Technologies

INTRODUCTION

This chapter discusses denial and survivability (D and S) technologies for a broad range of terrorist threat scenarios against assets and activities that are within the Army's mission area. Among the assets the Army will need to address for homeland security (HLS) D and S considerations are the following:

- Army bases, facilities, equipment, and troops;
- Assets the Army is temporarily responsible for safeguarding during times of threat; and
- Deploying forces in transit domestically.

"Denial of an attack," as used herein, refers to measures taken to prevent or otherwise thwart an intended terrorist attack, whether by preventing access through physical means (e.g., guards or barriers) or other means of interception (e.g., explosive detection, electronic surveillance). Survivability, in contrast, refers to measures taken to mitigate the effects of attack so as to reduce its effectiveness (e.g., by such means as structural hardening, protection of personnel, and duplication of resources). The elements of survivability also include the ability to absorb an attack with acceptable damage and casualties, redundancies that enable continued function after an attack, mitigation of the effects of the attack, and preparations for retaliation.

The line between D and S is not always a clear one. Consider as an example, building security. Denial relates to issues such as perimeter protection and entry control—denying the terrorist the ability to enter. Survivability relates to mitigating the effect of terrorist actions once the perimeter has been breached and entry obtained. Placement of barricades in such a way that a truck bomb produces inconsequential damage could be viewed as denial or survivability. Thus, it may not be useful to differentiate too finely between the two components when discussing applicable technologies.

The fixed infrastructure targets of primary interest to the Army are presumed to be installations, conventional military buildings either inside a base or standing alone (e.g., barracks, office buildings, and command and control (C2) centers), bridges, tunnels, and dams as well as special facilities such as nuclear power plants and critical Department of Defense (DoD)/Army assets (e.g., ports and airfields). Infrastructure targets also can include those that are primarily cybernetic, such as computer networks, communication systems, and C2 systems or supervisory control and data acquisition (SCADA) based systems such as military base power grids and water systems. Cyber issues will be addressed separately in the last section of this chapter.

Because many of these facilities are conventional, the technology that enhances their denial or survivability capabilities is equally applicable to civilian facilities and infrastructure. As was suggested in Chapter 1, technology transfer to the civilian sector will be necessary in order for the civilian sector to exploit Army technology.

The principal element of successful denial is good security, both physical and cyber. Security techniques and technologies that will satisfactorily perform Army HLS missions and protect against terrorist attacks will require "leap forward" capabilities. Simply doing more of the same or incrementally improving today's tool set will not result in affordable systems with acceptable performance. The Army must look for breakthrough technologies that not only enhance performance but also substantially reduce the resource demands of these functions.

PHYSICAL SECURITY

Security functions may provide the most leverage, both in terms of response options and resource savings. Security is also an area that could benefit enormously from new and innovative technology. Physical security includes activities at perimeters, gateways, and portals, as well as the detection of human agents. When the Army is deployed to protect a site in times of increased threat, the perimeters and portals may be temporary and not in optimal location or design, and portable or mobile systems may have to be used by the security force.[1] The desired attributes of the physical security functions are in listed Box 3-1.

---

[1]Providing adequate full-time protection for the dams, levees, bridges, tunnels, critical infrastructure, and Army structures will pose the challenge of balancing cost and public acceptability with available resources.

BOX 3-1
Desired Attributes for Physical Security

Perimeter Control

Boundary line monitoring. All-weather, day-night surveillance. Low cost, standoff sensing. Fenceless borders, low false and nuisance alarms. High detection rate. CBRN detection. Low manpower requirements. Air and ground threat detection. Difficult to spoof. Scalable. Secure, reliable communication to central command post. Sensor- and algorithmic-based assessment tools. Tools and equipment, such as robotic investigators, to assist human assessment.

Entry portal control. High throughput for authorized people and products. Rapid, positive ID of authorized personal. Forgery-resistant credentials. Rapid detection of threats in large vehicles (e.g. tank trucks, aircraft, or ships). CBRN detection and identification. Low risk to portal personal. Low manpower requirements. Rapid ID of nonauthorized attempts at entry. Safety setback for detected CBRN and LVBs. Deployable barriers.

Temporary perimeters. Rapidly deployable, flexible, scalable, all-weather, day-night surveillance systems. Simple to deploy with modular features. Fenceless borders, low false and nuisance alarms. High detection rate. CBRN detection. Low manpower requirements. Air and ground threat detection. Difficult to spoof. Sensor- and algorithmic-based assessment tools. Tools and equipment, such as robotic investigators, to assist human assessment.

Building and Facility Control

External protection. Access control systems that efficiently allow access only to authorized personnel. CBRN detection, neutralize/destroy integrated with HVAC technologies. Alarm systems integrated with local emergency response network.

Forces in Transit

Mobile protection. Vehicle-mounted area detection of CBRN and LVB threats.

NOTE: CBRN, chemical, biological, radiological, and nuclear; ID, identification; LVB, large-vehicle bomb; HVAC, heating, ventilation, and air conditioning.

The technology needs for physical security are very broad. Improved sensors are key to solving many of the problems identified here and are broadly described in Chapter 2.

New algorithms and techniques must be developed to allow rapid and faultless assessment of information about individuals attempting to gain access, material that is to be introduced into the facility, and detectors signaling a threat. Advances in data mining and cogitative modeling are essential. Tools to quickly identify unknown, unauthorized individuals using national law enforcement and intelligence databases need to be deployed to where the identification must take place. There is a need for an ability to search an integrated, seamless, real-time watch list. Such a capability does not now exist across all the relevant departments—e.g., Customs, the Immigration and Naturalization Service, the Federal Bureau of Investigation, the Central Intelligence Agency, and law enforcement at the state and local levels. These assessments and identifications must be made in an environment that protects our forces on the perimeter. This will require new concepts in perimeter and portal management and staffing. Table 3-1 identifies some of the technology challenges inherent in this task.

SURVIVABLE STRUCTURES

Blast Mitigation

Explosive threats against conventional buildings of direct interest to the Army may range from small 1- or 2-lb explosives packaged in letter bombs or pipe bombs, to hundreds of pounds of explosives contained in cars, to thousands of pounds of trinitrotoluene (TNT) equivalent charge carried by large trucks, trains, or dockside ships.

A bomb explosion in or near a building can have catastrophic effects, destroying or severely damaging portions of the building's external and internal structural framework, collapsing walls, blowing out large expanses of windows, and shutting down critical fire- and life-safety systems, such as fire detection and suppression, ventilation, light, water, sewage, and power.

Damage to a Building's Structure

Recent terrorist attacks against commercial buildings dramatically illustrate the influence of bomb placement and building design on the nature and extent of direct structural damage. Detonation of weapons inside or outside these buildings results in air-blast loadings that disintegrate the relatively weak front face slabs and curtain walls and/or damage columns through direct loading and partial transfer of the loads from the weak slabs. Failure of columns or load-bearing walls due to a combination of lateral air-blast loading plus axial gravitational forces from the weight of the structure above it may result in progressive collapse of the building or portions of it.

Notable examples of the damage potential of external explosions against multistory buildings that led to progressive failure are the 1995 bombing of the Alfred P. Murrah Federal Building in Oklahoma City (the largest such terrorist attack in the United States up to that time caused 168 fatalities, numerous injuries, and an estimated $50 million in damage to about 75 buildings in the area) and the devastating 1994 car bomb attack against the Jewish Community Center in Buenos Aires (a masonry load-bearing wall building whose collapse killed 87 people and injured 200 others). By way of contrast, a similar attack in 1992 against a multistory office building of more modern concrete column and slab

TABLE 3-1 Technologies for Physical Security

Function	Task	Technology	Characteristics	Availability[a] (R, N, F)	Priority for Army S&T[b]	Multiuse[c] (H, O, C)
Perimeter control	Modeling of perimeter control system	Advanced decision theory; high fidelity, interactive virtual reality; full feature models	High-fidelity, flexible simulation of system performance	N	Medium	H, C
	Rapid detection of LVBs	X-ray assessment, swimming sensors	Detection of LVB hidden in tank trucks and other normal base traffic	N, F	High	H, O
	High-performance fenceless perimeters	Laser interrogators; microwave networks; robotic rovers	Perimeters with high probability of detection, low false-alarm rates for full threat spectrum	N, F	Medium	H, O, C
	Tool to assess alarms	Robotic investigators; cognitive networks	Mobile, low-manpower requirements, multisensor, networked	F	Medium	H, O, C
	High performance credentials	Smart ID with bioinformation; ID tracking with area authorization	Positively ID authorized personnel	N, F	Medium	H, O, C
	Biometric recognition	Iris ID, liveness tests, auto DNA ID	Positively ID authorized personnel	N, F	Medium	H, O, C
	ID nonauthorized visitors	3-D facial recognition; auto DNA matching	Link to national database to ID attempted unauthorized entry	F	Medium	H, O, C
	Protection of perimeter guard forces from LVBs and other WMD devices	Remote interrogation, positive barriers	Remote assessment capability and blast protection	F	Medium	H, O

	Deployable perimeter control system	Unattended sensor networks, advanced power sources, C2 and secure communication, low-power sensing elements	Modular, robust deployable perimeter control system for use by Army in high-threat situations	N, F	High	H, O
	Mobile perimeter system to include CBRN and LVB detectors for force protection during deployment	C2 and secure communications, situational awareness tools, area sensors	Vehicle-mounted, networked, detection systems to protect forces during transit to ports and airfields	F	High	H, O
Building and facility control	CBRN detection and remediation	See discussion in Chapter 2	Buildingwide system to detect, prevent wide dispersal, and mitigate or destroy CBRN agents. Secure communication to civilian emergency responders	N, F	Medium	H, O
	Automatic, high-confidence access control	Smart ID with bioinformation, ID tracking with area authorization, iris ID, liveness tests, auto DNA ID	Rapid and faultless ID of authorized individuals and detection of dangerous articles	F	High	H, O, C

NOTE: LVB, large vehicle bomb; ID, identification; 3-D, three-dimensional; C2, command and control; CBRN, chemical, biological, radiological, and nuclear; TRL, technology readiness level.

[a]Availability: R, ready (TRL 8-9); N, near-term (TRL 4-7); F, far-term (TRL 1-3).

[b]Priority for Army S&T (investment): low, someone else has mission or technology is ready and available; medium, useful but of limited impact and some investment needed; high, very important, no one else working on it, considerable investment needed.

[c]Multiuse: H, Army homeland security; O, Objective Force; C, civilian (first responders and others).

construction at St. Mary Axe in London produced relatively minor structural damage but extensive glass damage[2] (NRC, 1995).

In addition to the direct effects of an explosion or impact, the causal event may initiate a fire that can be fed by existing materials in the building. These fires can reduce the strength of structural steel by 50 percent if they reach temperatures of 500°C and to near zero if the temperatures reach 1000°C (NRC, 2002). As noted in Chapter 8 of the NRC report, "columns, floor diaphragms, and connections between the columns and floor joists are the vulnerable members" (NRC, 2002). This weakening may occur despite the presence of fireproofing, because the force of the explosion or impact and the debris from it may strip the fireproofing from the structural elements and assemblies. In addition, the fireproofing may have been applied improperly or removed over the course of time. Current building codes[3] do not consider the combined effects of fire and impact or blast on the integrity of the fire protection system. Generally, normal-strength concrete members demonstrate good performance under fire exposure. However, low-strength concrete and high-strength concrete may not perform as well under severe fire conditions (FEMA, 2002).

Damage to Building Subsystems

Certain building subsystems, if lost, render the building unable to protect the occupants or assist in their survival and otherwise make the building uninhabitable or unusable. Typical of these subsystems are fire-detection and fire-suppression systems; water and sewer service, including sanitation; means of egress, including corridors, stairs, lobbies, and exit doors; elevators; primary and emergency electrical systems; and rescue operation systems, including voice and data communications, ventilation, and smoke control. A bomb detonated inside a building's parking garage can cause serious damage to building subsystems simply because several critical subsystems typically originate there, along with much of the control and distribution equipment. A garage-level detonation has a significant potential for fire and smoke production because the parked vehicles contain large amounts of combustible materials. Also, the fire-suppression system would likely be made inoperable, since it is exposed and very fragile.

The 1993 World Trade Center bombing was, unfortunately, a good example of these observations: Extensive damage occurred to communications, life-safety, electrical, and mechanical systems; the emergency generator plant shut down

---

[2]The offices were unoccupied at the time of the explosion (around midnight). It is thought that extensive injuries would have occurred to occupants had the bomb been detonated during working hours (NRC, 1995).

[3]The building codes are referenced in Chapter 1 of FEMA (2002), and the fire protection codes are referenced in Appendix C of that document.

because of loss of cooling water; the elevator and stair shafts were breached; smoke from burning automobiles on the parking levels was forced up the shafts of both towers; and the underground tower's operations control center was put out of commission, leaving building occupants without important information (NRC, 1995).

Hazards to People

Injuries and loss of life can result directly from the explosion of a bomb. Blast pressure, impact of high-speed glass fragments or other structural debris, collapse of structural members, fire and smoke inhalation, or a variety of other causes associated with the general confusion that may follow an explosion and a possibly prolonged evacuation period can all contribute to casualties. After entrapment in collapsed building spaces, the next most serious source of injuries is missile penetration or smoke inhalation. Additionally, toxic gases and dusts from conventional blasts may become entrapped in the urban environment for days or weeks. This form of pollution may be another target for monitors (and sensors). The harmful effects of dusts, vapors, and gases on an urban civilian population could be quite serious.

The breaching of elevator and stairwell doors (more likely from street-level explosions) allows smoke to migrate upward into the building, carried by the building's stack effect during winter months. Elevators are likely to be occupied throughout the day, and persons may be trapped within them, as a result of either damage to the elevator shaft or hoists or the loss of power or controls. In the 1993 World Trade Center bombing, the north tower air locks were destroyed, and smoke and dust-laden air were forced to the upper floors, accounting for most of the more than 1,000 personal injuries (NRC, 1995).

Conclusion 3-1. The current database describing injuries and fatalities due to blast-related terrorist activities is sparse.

Recommendation 3-1. To gather valuable and perishable medical and other forensic data, the Army should support the establishment of rapid response data-gathering teams to investigate bombing attacks that may occur in the future. The data collected by these teams should be integrated with information from past events and made available to researchers and practitioners in emergency medicine, injury epidemiology, search and rescue, architecture, and engineering.

Technology for Blast Mitigation

The trend in civilian building design for the last 50 years has been toward the use of lighter but stronger materials. This has led to more economical buildings, with the structure accounting for less of the floor area and lower first costs. At the same time, engineers developed a better understanding of building performance when a structure is subjected to dynamic horizontal and vertical forces associated with wind and earthquake. Seismic design calls for the building to possess adequate strength (force- and ductility-resistance characteristics) to resist repetitive seismic motions in a manner that protects human lives and leaves the building usable or, at worst, with damage that is easily repairable.

The dynamic loading on buildings caused by explosions differs in important respects from dynamic loads caused by earthquake and wind.[4] The latter loads are of relatively low intensity, long duration (seconds to minutes), and essentially oscillatory (periodic in nature). Explosive loads, by comparison, are extremely large initially, act for very short durations of time (milliseconds), and are non-oscillatory (aperiodic). To effectively resist large, short-duration explosive loads localized in lower levels, characteristic of terrorist bombings, the mass of the lower levels of a structure should be increased. This goal is generally in keeping with seismic requirements, which call for significant strength in the lower levels. In other respects, however, the two design approaches differ considerably.

Design of New Facilities

A series of manuals exists for the design of new facilities subjected to the kinds of threats described above.[5] These manuals include charts and/or fast-running computer codes to forecast the threat environments, including blast, fragments, and ground shock.

Retrofit of Existing Facilities

The retrofit of existing buildings presents a different challenge to the designer because of the many constraints imposed by the need to retain a building's functionality while retrofitting is occurring. This need imposes limitations on volume and configuration available for retrofit approaches and imposes addi-

---

[4] A discussion of the design and behavior of structural components typically used in modern civilian buildings subjected to a transient blast-wave form is contained in Chapter 4 of *Structural Design for Physical Security* (ASCE, 1999).

[5] USACE TM 5-855-1, *Fundamentals of Protective Design for Conventional Weapons*, 1986; USACE TM 5-1300, *Structures to Resist the Effects of Accidental Explosions*, 1990; USACE TM 5-853, *Security Engineering*, 1993; and, most recently, the new joint services DAHS/CWE manual, *The Design and Analysis of Hardened Structures to Conventional Weapons Effects*, 1995, which is computerized and interactive. USACE TM 5-853 provides a systematic methodology to analyze "aggressor threats and tactics," including a system for rating potential risks and developing appropriate responses. It discusses various design options to a limited degree, but the planning techniques are strongest in the area of supporting access control to the facility.

tional hazards that must be addressed, e.g., the retrofit of masonry and brick must address the containment of projectiles of these materials created by an explosion. In addition, standoff—that is, the distance between the building and a potential device—may be minimal or nonexistent.

Standard retrofit procedures consider the introduction of additional strength, ductility, redundancy, and mass and the replacement of weak structural components. They can include the enhancement of support conditions through better connections, span reduction, the strengthening of exterior facades such as curtain walls, the strengthening of interior partitions, and the installation of windows and doors with better blast resistance and seals. Many of these options are presented in *Structural Design for Physical Security* (ASCE, 1999).

Chemical, Biological, and Radiological Threats

Military and conventional buildings are susceptible to chemical, biological, and radiological (CBR) attacks by terrorists through their heating, ventilation, and air conditioning (HVAC) systems. The effectiveness of such attacks can be greatly reduced by incorporating a building automation system designed to manage specific threats and scenarios. Such systems can include detection, isolation, neutralization, and, possibly, decontamination. The HVAC systems can be improved and integrated with architectural/civil design features for both new buildings and retrofits to gain more effective resistance to CBR attacks. New developments in real-time monitoring devices, filtration and chemistry for detection, neutralization, and decontamination of CBR agents can be combined with modeling and simulation tools to isolate and manage the terrorist threat. Some simple steps that can be taken for existing buildings are presented in NIOSH (2002).

This is the focus of a new DARPA research program for "immune buildings," which seeks to modify and augment the building infrastructure to make buildings far less attractive targets for attack by airborne or aerosolized chemical or biological warfare agents. The program has three goals: to protect the human inhabitants of such buildings in the event of an attack; to restore the building to full function as quickly as possible after the attack; and to preserve forensic evidence for attribution and retaliation. Release of biological agents inside a building is the most challenging threat, as it requires a rapid response to stop or neutralize the agents before they affect humans. The utilization of large-volume, nonthermal diffused plasmas that can be generated at ambient pressure for contaminant conversion, along with existing or improved building filtration technology, looks promising (DARPA, 2002).

Conclusion 3-2. Heating, ventilation, and air conditioning systems can be improved and integrated with architectural/civil design features for both new buildings and retrofits to provide better resistance to chemical, biological, and radiological attacks.

Recommendation 3-2. The Army should monitor and integrate new heating, ventilation, and air-conditioning technologies developed by the Defense Advanced Research Projects Agency and other organizations into building and infrastructure design and retrofit guidelines. These technologies include detection, neutralization, filtration, and active ventilation defenses.

Technology Gaps

It might appear from the above discussion that ample information is available for the architect/engineer to provide blast-mitigation designs for both new and retrofit structures. Unfortunately, this is not the case, because much of the required information either is not directly applicable to the construction of modern commercial buildings or is inaccessible to most practitioners in the commercial building industry and difficult, if not impossible, to use. A 1995 report makes clear that translating blast-effects research into practice will be a major undertaking. It is in any case an undertaking that the committee believes the U.S. Army Corps of Engineers (USACE) is uniquely positioned to lead (NRC, 1995).

Table 3-2 lists the technologies required to protect people and buildings from terrorist threats in both new and existing structures.

Current Research and Development Efforts—Leveraging the Army's Contribution

The Technical Support Working Group (TSWG)/Defense Threat Reduction Agency (DTRA) Blast Mitigation for Structures Program is a focused and valuable program of research, testing, engineering analysis, and computational modeling to supplement existing knowledge on blast effects and blast-resistant design and construction. However, the full benefits of the program will be realized only if the results are widely disseminated and necessary improvements implemented. The USACE is the logical choice to facilitate a continuing technology development and transfer effort because of its long involvement in both research and development, and in developing design guidance for architects and engineers.

The USACE and its Omaha District Protective Design Center are also participating in DARPA's Immune Building Program. There is an opportunity for the USACE to play a more active role in the demonstration phase of this program and to be a principal source of technology transfer to the building industry.

Physical Security Summary

As was noted earlier, blast-hardening technologies and design principles developed by the Army and other DoD components for military purposes are generally relevant for federal force protection and civilian design practice. However, because the knowledge base is incomplete, this information must be adapted

TABLE 3-2 Technologies for Blast Resistance of Building Structures for New and Retrofit Construction

Function	Technology	Characteristics	Availability[a] (R, N, F)	Priority for Army S&T[b]	Multiuse[c] (H, O, C)
Environment definition	Prediction of blast and impact loads on and in buildings, bridges, dams, etc.	More effective designs for better defined loads	N, F	High	H, O, C
	Effects of barriers on blast mitigation	Reduce loads on buildings	R, N	Medium	H, C
	Characterization of new explosives and gas deflagration	Model new loads on structures	N	Medium	H, O, C
	Characterization of debris	Design for fragment impact and human injury	N, F	Medium	H, C
	CBRN and fire propagation in buildings and tunnels	Evaluate effects on structural integrity, equipment, and personnel	N	Medium	H, C
Structural strengthening	Column blast design and retrofit using conventional and new materials	Upgrade existing and new designs more efficiently	R, N	Medium	H, O, C
	Slab retrofit	Same as above	R, N	Medium	H, O, C
	Wall retrofit, including load-bearing masonry walls	Same as above with new complexity due to projectile resistance requirement due to masonry breakup	R, N	Medium	H, O, C
	Connection details for steel and concrete structures (new and retrofit construction)	Upgrade current approaches for dynamic environments and material behavior	N	High	H, O, C

Continues

TABLE 3-2 Continued

Function	Technology	Characteristics	Availability[a] (R, N, F)	Priority for Army S&T[b]	Multiuse[c] (H, O, C)
	Methodology to prevent/ evaluate potential for progressive collapse	Improve capabilities to evaluate collapse rapidly; new approaches for design	N	High	H, O, C
Windows and curtain walls	Blast-resistant window concepts, including new glazing-to-frame connections	Look at glazing and frame as system	N	High	H, O, C
	Curtain wall concepts for energy absorption	Explore system approach for windows and frames	N	Medium	H, C
Materials research	Blast-resistant tempered and laminated glass (stiffness, strength enhancement, ductility)	Little material characterization for constitutive modeling and evaluating new concepts	F	High (+ university, industry)[d]	H, C
	Materials testing and analysis of fire resistance	Ability of insulation and insulated structural members and connections to survive blast environment and specified duration of fire	N	Medium	H, C
	High-temperature properties of building materials, including insulation and structural materials.	Ability to evaluate collapse of structures in thermal environments	F	Low (+ university, industry)[d]	H, C
	Special blast- and fire-resistant materials: Kevlar, LINEX, and other textiles; graphite epoxy and other composite materials for use in retrofit designs	Material property characterization in blast and thermal environments lacking	N	Medium	H, O, C

	Energy-absorbing materials for barriers	New, efficient concepts that do not generate hazardous projectiles	N	Medium	H, O, C
Design guides and computer codes	First-principles analysis techniques to supplement experimental databases for design of windows and structural component retrofits	For design and evaluation of new concepts and configurations; supplement experimental database	N	High	H, O, C
	Optimization software for new designs involving multihazard scenarios, including seismic, wind, tornado, blast, fire, and chem/bio threats	Develop mathematical tool to achieve a balanced and economical design for multiple hazards	N, F	Low (+ university, industry)[d]	H, C
	Software to include new test and analysis data and techniques for design and retrofit of structures in blast environments	Upgrade existing packages to include new data and methodology	R, N	High	H, O, C
	Integration of performance standards with building codes from a multihazard perspective	Incorporate and integrate new design standards and procedures in official codes for multihazard environments	N, F	High	H, O, C

NOTE: CBRN, chemical, biological, radiological, and nuclear; TRL, technology readiness level.

[a]Availability: R, ready (TRL 8-9); N, near-term (TRL 4-7); F, far-term (TRL 1-3).

[b]Priority for Army S&T (investment): low, someone else has mission or technology is ready and available; medium, useful but of limited impact and some investment needed; high, very important, no one else working on it, considerable investment needed.

[c]Multiuse: H, Army homeland security; O, Objective Force; C, civilian (first responders and others).

[d]Parenthetical entries suggest that participation by universities and/or industry should be especially sought because their technology, understanding, experience, and/or scientific capabilities in these areas are advanced, their databases are useful, and their participation would provide new insight and/or information to the program and shorten the time frame for development.

and expanded to be more specifically usable by and accessible to civilian architects and engineers.

The ongoing TSWG/DTRA Blast Mitigation for Structures Program, in which the USACE Environmental Research and Development Center is a major participant, is a natural vehicle for such technology development and transfer. This should include research and testing of common building materials, assemblies, equipment, and associated designs applicable to the blast-resistant design of critical nonstructural, life-safety, building subsystems. Techniques and products for the retrofit of existing buildings to protect against multiple hazards such as earthquakes, extreme wind events, fire, and flood, as well as blast effects, should be developed. Implementation of blast-mitigation measures should utilize established risk management principles that integrate security and natural hazard mitigation objectives with new technologies and should be based on building mission, defined threat, acceptable risk, and available resources.

Glass material properties must be characterized in a form suitable for modeling and simulation in order to be able to predict the response and failure of windows subjected to blast loading. Research in this area is being conducted by universities under government and private sponsorship. Universities also conduct research on blast and impact loading and the response of structures.

Conclusion 3-3. Research currently being conducted by universities in window/glass behavior and structural response through failure in dynamic environments can help to improve the blast resistance of key structures.

Recommendation 3-3. The Army should continue to survey and evaluate relevant ongoing university research with the objective of identifying and synthesizing technology that could improve the performance of buildings in a blast environment, and it should also consider inviting universities to directly participate in the research effort.

INFORMATION SECURITY AND CYBER ISSUES

The committee uses the word "cyber" to refer to any activities related to the computer and communications (C&C) infrastructure, including the information stored in and/or being transmitted by the systems. This infrastructure is rapidly becoming ubiquitous in all aspects of daily life as well as for first responders: C2 systems are often based on it, medical information systems and financial systems are based on it, other infrastructures such as water and energy are based on it through SCADA, and it is being used in newer versions of almost everything electronic, such as monitoring systems, from perimeter control to baby watching. One has only to read the popular press to hear of proposals to give an Internet Protocol (IP) address to every device from a toaster on up to a washing machine to appreciate the drive to interconnect everything. At the same time there is a movement to make almost all devices software-based so that updates can be downloaded over the connected network to provide the flexibility for future changes.

The C&C infrastructure can be compromised in several ways, principally the following:

- An insider making use of authorized access,
- Unauthorized access via direct tapping into the physical facility,
- Unauthorized access via valid network connections and security flaws in the system, or
- Denial-of-service attacks.

Protection against the first two threats is based on physical security of the facilities and control of personnel. These are common security issues where countermeasures have been well studied, so the committee will not discuss them further here. However, even if the perimeter or the hardware is breached, damage must be contained. In the cyber context, this means that gaining access to one subsystem within a security perimeter must not automatically grant access to other subsystems.

Range of Threats

There are three primary objectives of a cyberattack:[6]

- Destroy or change data within the system itself,
- Take control of systems controlled by the C&C system, or
- Deny the user effective use of the system.

Future terrorist incidents in the United States could attempt any of these. Institutions from financial to medical would have serious problems in the event of massive loss of data or of reasonably rapid network access to it, but neither protection against this nor remediation if it happens fall within the Army's jurisdiction. (However, the Army does need to protect its own systems from such attacks.)

When a computer system with control functions is compromised by attack, the community may face problems as the controlled entity fails to operate cor-

[6]Attacks by hackers merely to prove they can do it by making annoying but inconsequential changes to the system are not discussed. It should be recognized that many of these hacker attacks are against that part of the network that is designed to be public—namely, the Web site. While it is desirable to keep those pages secure against unauthorized change, the level of security that can be achieved is necessarily lower than that which can be applied to nonpublic information.

rectly. This could happen whether the attacker is actually able to take control of the system and redirect it or is just able to interfere with its correct operation.

A denial-of-service attack is the overloading of a C&C system with superficially legitimate service requests via the network. It does not require any security flaws or other break-in technology, but such an attack could be used to deny or corrupt important services as a preliminary or follow-up to a physical attack. For example, if an emergency response group relied on public Web-based data access for its functionality, it could be susceptible to a denial-of-service attack. Non-public systems would require the exploitation of a security flaw to deny service.

Mitigation Technologies

The best defense is to physically isolate an important network from the public network. However, it is dangerous to assume that this will resolve all problems. The additional functionality that can be obtained by interconnecting units frequently leads to the addition of network interconnections or unauthorized access. For example, the committee learned of executives who connect their office phones to computer modems so they can work from home, thereby providing an opportunity for access by others.[7] Some systems provide for progressive shutdown of connections as the perceived threat level increases. However, it should be realized that certain forms of cyberattack can be preplanted before there is evidence of a raised threat level and left to activate automatically later. For this reason, it is important to defend against threats to networked systems.

The primary threat to networked computer systems comes by way of security flaws in the system that allow remote access to unauthorized users. It is important to realize that C&C systems are sufficiently complex that it is highly unlikely a system can be designed that does not contain any security flaws. One must therefore accept the fact that providing security is an ongoing operation and cannot be built in with 100 percent certainty. Hence the initial design must pay great attention not only to achieving a high initial level of security but also to locating and correcting flaws during the lifetime of the system. It is also important to realize that a C&C system is not a static design but typically evolves as new or changed functionality is introduced. Such changes often introduce new security flaws.

Large organizations are often tempted to custom design their own systems because they believe that their needs are significantly different and because they believe they can achieve greater efficiency by dropping system requirements they do not have, at least not at the time of design. For general-purpose systems this is not only a false economy—the design costs are such that because of the rate of

[7]Herb Lin, Computer Science and Technology Board, National Research Council, briefing to the committee on July 24, 2002.

change in the field, the organization will soon be left with an out-of-date design that runs on out-of-date hardware—but it is also an invitation to security disasters. While it may seem that the use of commercial off-the-shelf systems means that more people will know where the flaws are, it also means that vastly more people are busy looking for those flaws and bringing their skills to the task of fixing them. Clearly the Army must work with other interested parties to achieve the maximum level of protection.

Document P of the National Infrastructure Assurance Plan (*Planning Guidance to Assist in the Development of the Response Functional Plan*), notes as follows:

> Resolving the inherent overlap of responsibilities and capabilities while defining the roles of FEMA v. FBI (including the Cyber Emergency Support Team) in developing this plan will be a critical step in implementing this plan. Additionally, other government departments (e.g., Defense) are developing cyber response capabilities. There will be a need to share best practices among these efforts and clarify the responsibility across the government (NIPC, DoD, etc.) and with the private sector (DoC, 1998).

From this the committee draws a conclusion and two recommendations:

Conclusion 3-4. As the Army becomes more dependent on computer-based systems, cybersecurity becomes more of an issue.

Recommendation 3-4a. The Army should partner with other agencies and the commercial sector to develop and adopt the appropriate tools and protocols for the protection of its own computer and communication systems.

Recommendation 3-4b. The Army should continue to review its cybersecurity procedures to assure that the best practices from the community are adopted on an ongoing basis.

The Army does not currently have a direct role in denial/survivability for any non-Army C&C systems, but should coordinate with those agencies that do.

Survivability

The Army must not only be concerned with the survivability of its own systems in the event of an attack, it needs to be concerned with the survivability of systems over which it has no or little control prior to the attack—or even, perhaps, after the attack, since if it is called on to provide support, it will need to establish links between its units and civilian responders. The characteristics of the systems are shown in Table 3-3.

TABLE 3-3 Technologies for Cybersecurity

Function	Task	Technology	Characteristics	Availability[a] (R, N, F)	Priority for Army S&T[b]	Multiuse[c] (H, O, C)
Continuous monitoring of Army C&C systems	Cyber perimeter protection	Firewalls	Limits traffic to trusted sites.	R[d]	[e]	
		Virus scanning	Checks incoming files for viruses, worms.	R[d]	[e]	
	Ad hoc mobile C&C networks to rapidly reconfigure systems	IP version 6	Imparts ability to dynamically reconfigure networks as systems arrive, leave, are destroyed, etc. Low power and security are issues.	N	High	H, O, C
Physical monitoring	Sensor networks	Developing	Similar to above for sensors. Will couple to C&C networks.	N	Medium	H, O, C
		Data fusion	Combines multiple sensor information.	N-F	Medium	H, O, C

Security of rapidly deployed ad hoc networks[f]	Developing	Avoids enemy intrusions, guarantees functionality.	F	High	H, O
Rapid emergency deployment of C&C capacity[f]	Various	Gives ability to provide alternative C&C after a disaster.	N	High	H, O
Ad hoc interoperability[f]	IP version 6 for networks, universal radio, etc.	Allows the Army systems to interoperate with other emergency services.	N	High	H, O

NOTE: C&C, computers and communication; IP, Internet Protocol; TRL, technology readiness level.

[a]Availability: R, ready (TRL 8-9); N, near-term (TRL 4-7); F, far-term (TRL 1-3).

[b]Priority for Army S&T (investment): low, someone else has mission or technology is ready and available; medium, useful but of limited impact and some investment needed; high, very important, no one else working on it, considerable investment needed.

[c]Multiuse: H, Army homeland security; O, Objective Force; C, civilian (first responders and others).

[d]Technology available now is continuously being updated, and the Army must stay current.

[e]This should not be viewed as an S&T investment, but as necessary system administration. The Army should continuously adopt best practices from the community.

[f]These are adapted from Table 5.1 of *Making the Nation Safer* (NRC, 2002).

From considering the aftermath of the attacks of September 11, 2001, one can conclude as follows:

Conclusion 3-5. Even if the attack does not directly inflict physical or cyberdamage on computer and communication systems, the public systems may become overloaded. Since the first responders often use components of public systems, command, control, communications, computers, intelligence, surveillance , and reconnaissance (C4ISR) may be a significant problem in the aftermath.

The executive summary of the Hart-Rudman phase 3 report states

> We urge, in particular, that the National Guard be given homeland security as a primary mission, as the U.S. Constitution itself ordains. The National Guard should be reorganized, trained, and equipped to undertake that mission. (Hart and Rudman, 2001)

In light of the aforementioned conclusion, the committee asserts as follows:

Recommendation 3-5a. Whether through the Army National Guard or active or reserve Army units, the Army should play a major role in providing emergency command, control, communications, computers, intelligence, surveillance, and reconnaissance (C4ISR) in the event of a major natural or terrorism disaster because it has both the skill set and the equipment to provide such services in hostile environments.

Recommendation 3-5b. Equipment and trained personnel should be available to provide vital information and communications for interoperable command, control, communications, computers, intelligence, surveillance, and reconnaissance (C4ISR) in the case that civilian systems are seriously impaired in an emergency event.

In some situations an impairment would occur simply because existing public facilities would be overused by concerned citizens. In that case, it might be desirable for the Army to provide alternative systems for emergency services.

The Army already has a strong interest in and need for mobile battlefield networks. One such system (MOSAIC) is currently an advanced technology demonstration (see Chapter 4.) These networks differ from civilian and most other networks in being ad hoc, since there can be no fixed hubs on a moving battlefield. Such systems would be very useful after an incident if there is significant disruption to the standard communications in the area (network and voice). For this to happen, Army systems must be interoperable with current civilian technology. Enhancements to existing Army systems should reflect the need for multiuse capabilities, and new battlefield systems should be designed with both civilian interface and domestic and foreign missions in mind.

SUMMARY

Denial and survivability (D and S) issues will affect a very broad range of activities that are within the Army's mission area. The assets that the Army will need to counter the events that might arise during this period may, in some instances, differ quite dramatically from those required in a conventional wartime environment. However, whether the tools relate to the built environment or the cyber environment, the Army must prepare.

REFERENCES

ASCE (American Society of Civil Engineers). 1999. Structural Design for Physical Security: State of the Practice. Reston, Va.: ASCE.

DARPA (Defense Advanced Research Projects Agency). 2002. Immune Building Program. Available online at <http://www.darpa.mil/spo/programs/immunebuilding.htm>. Accessed on October 2, 2002.

DoC (Department of Commerce). 1998. National Infrastructure Assurance Plan, Document P: Planning Guidance to Assist in the Development of the Response Functional Plan. Washington, D.C.: DoC Critical Infrastructure Assurance Office.

FEMA (Federal Emergency Management Agency). 2002. World Trade Center Building Performance Study: Data Collection, Preliminary Observations, and Recommendations, FEMA 403, May. Available online at <http://www.fema.gov/library/wtcstudy.shtm>. Accessed on October 2, 2002.

Hart, G., and W. Rudman. 2001. Road Map for National Security: Imperative for Change: The Phase III Report of the U.S. Commission on National Security/21st Century. Available online at <http://www.nssg.gov/PhaseIIIFR.pdf>. Accessed on October 3, 2002.

NIOSH (National Institute for Occupational Safety and Health). 2002. Guidance for Protecting Building Environments from Airborne Chemical, Biological or Radiological Attacks. Available online at <http://www.cdc.gov/niosh/bldvent/2002-139.html>. Accessed on October 7, 2002.

NRC (National Research Council). 1995. Protecting Buildings from Bomb Damage: Transfer of Blast-Effects Mitigation Technologies from Military to Civilian Applications. Washington, D.C.: National Academy Press.

NRC. 2002. Making the Nation Safer: The Role of Science and Technology in Countering Terrorism. Washington, D.C.: National Academy Press.

4

Recovery and Consequence Management Technologies

INTRODUCTION

The purpose of this chapter is to identify science and technology (S&T) initiatives that will enhance the ability of the Army to accomplish its emerging mission requirements for homeland security (HLS). This chapter is focused on the recovery and consequence management (R and CM) functions.

Generally, recovery is viewed as a local and private sector responsibility. However, in the case of terrorist acts using weapons of mass destruction (WMD), or significant cyberattacks on the nation's critical infrastructure, the damage may exceed the capacity of local agencies and the private sector that owns and operates the critical infrastructure. In this situation, the nation's military, most likely the Army, would be called upon. Compounding the seriousness of the situation is the fact that R and CM mission activities may need to be conducted simultaneously with missions to protect the critical infrastructure or to conduct contingency operations overseas.

Consequence management (CM) is concerned with minimizing the damage resulting from a disruptive event (White House, 1998). CM is often conducted in conjunction with crisis management activities. Crisis management is a law enforcement mission aimed at early detection, prevention, and elimination of the cause of a disruption as quickly as possible (White House, 1998). There is an overlap in the crisis management and the CM missions. Mitigating the effects of a terrorist event and restoring public order and essential services is the principal objective of CM.

NEW MISSION CHALLENGES

The new HLS mission requirements are still under development, but a review of the likely challenges and threats can provide insights into the new missions and capabilities that will be necessary. The need to assist authorities in restoring order, overcoming the effects of physical damage, and beginning the road to recovery will remain. The Army has demonstrated its ability to meet these challenges. It has been fortunate that the training, equipment, and organizational constructs developed for wartime mission and contingency operations have met these challenges, by and large. However, the security environment of the 21st century poses new demands that call for new capabilities. The work that the Army has accomplished as part of the Objective Force in developing adaptive force packaging will be important in providing the right types and numbers of forces to meet the new challenges associated with HLS missions.

The effects of chemical, biological, radiological, nuclear, and high explosive (CBRNE) weapons can far exceed the effects (in time and scale) of even the largest natural disasters. These weapons can cause large numbers of casualties that are beyond the capacity of the civilian medical care system to address. Compounding the effects of physical destruction, chaos, and casualties, such weapons leave behind chemical, biological, and radiological contamination that can continue to cause death and disease and must be contained and cleaned up before public order is restored and recovery is initiated. In addition to large-scale R and CM operations for WMD, Army forces may be called upon to provide R and CM activities for a cyberattack on the nation's critical infrastructure. Such an attack could deny power and communications to wide areas, cause massive disruption in the nation's transportation and financial systems, and deny essential government services. Multiple events where WMD are employed against the nation, combined with cyberattacks against the critical infrastructure, could be even more challenging. Without a clearly defined mission for the Army, the committee postulates that it would participate in many of the following tasks as part of the R and CM phase.

Postulated Tasks

Initial Response

- Deploy forces.
- Protect responding forces.
- Identify the on-scene commander.
- Establish an interoperable command, control, communications, computers, intelligence, surveillance, and reconnaissance (C4ISR) system with existing civilian and military assets.

- Assess in real time the extent of the physical damage, casualties, and the enduring level of contamination and risk of disease transmission.
- Establish quarantine zones, safe areas, and perimeter control of movement.
- Triage and treat the injured.
- Preserve forensic information.
- Establish an information clearinghouse.

Containment

- Expand the area of control and model/predict moving boundaries.
- Isolate secondary threats (gas mains, electrical service, stability of damaged infrastructures and buildings).
- Restore or replace (substitute) infrastructure critical to containment.

Near-Term Recovery

- Keep the population informed.
- Eliminate/control ongoing immediate threat (contain the effects of WMD).
- Expand the treatment of casualties (begin stress management, including for military responders).
- Rescue, protect, evacuate, and track civilians.
- Assure food and water safety.
- Provide shelter, food, and support for personnel in the affected areas.
- Establish and validate the census of people and resources.
- Determine, marshal, and deploy forces required for long-term operations.

Restoration of Normalcy

- Decontaminate the effects of WMD.
- Consolidate deployment of forces.
- Establish or become part of an interoperable C4ISR system.
- Assess in real time the extent of the physical damage, casualties, and the enduring level of contamination and risk of disease transmission.
- Restore public order and essential services.
- Protect consequence management personnel.
- Move essential provisions.
- Establish quarantine zones and safe areas.
- Treat mass casualties.
- Secure the area and communicate the area of control.
- Reestablish lost essential facilities and infrastructure.
- Restore the physical infrastructure.

The Army will not be called upon to conduct all of these missions by itself but will support civil authorities. Numerous other agencies, including many in the private sector, will also have a significant role. However, in an event of national significance, the military may be called upon to take over where other institutions lack the capacity. Part of the Army's challenge will be to work in conjunction with the Northern Command (NORTHCOM) and the new Department of Homeland Security (DHS) to define the extent of potential missions prior to the occurrence of events that require large-scale consequence management.

REQUIRED TECHNOLOGIES AND CAPABILITIES

The infusion of new capabilities and technology will enhance the ability of the Army to conduct large-scale R and CM activities in conjunction with other agencies. The Army currently possesses significant capability to meet many of the challenges described in the preceding section. Through planning, organization, and training, the Army can satisfy other mission challenges as well. Many of the needed capabilities can be achieved as part of the Objective Force. Nevertheless, the Army will need to monitor developments to leverage promising technologies and to assure interoperability with the local, state, and federal agencies participating in the HLS mission. In the mission capabilities described above, several areas are ripe for exploitation by the Army and lend themselves to the application of Army S&T and apply to both HLS and the Objective Force mission. The areas of concentration include the following:

- Establishment of, or integration into, an interoperable C4ISR system;
- Real-time assessment of physical damage, casualties, and the enduring level of contamination;
- Force protection;
- Treatment of mass casualties; and
- Containment of and, later, decontamination of the effects of WMD.

The Army already has the capacity in other mission areas, provided that the appropriate doctrine is developed and that plans are established across the government and in NORTHCOM.

Interoperable Command, Control, Communications, Computer, Intelligence, Surveillance, and Reconnaissance System

Over the next few years, it is expected that the HLS organization will establish a national emergency response command and control (C2) system. Many different systems exist today across the numerous departments and agencies that are being blended into the DHS. Each system was created for (and is currently used for) a variety of purposes and missions. Very few are interoperable. Indeed, in crisis and consequence management incidents over the last decade, responders consistently report that an unwieldy number of different radios and wireless devices were needed to talk to the other participants. It is possible that lives were lost because the first responders were unable to communicate and share their situational awareness.

There is a strong need for an integrated system that allows the new HLS structure to conduct operations effectively; share a common operational picture built on a common database; provide multilevel security information to accommodate local, state, and federal needs; and facilitate real-time communications between these local, state, and federal entities.

The Army has already designed a mobile battlefield network system that might meet many of the DHS needs. As discussed briefly in Chapter 3, the Multifunctional On-the-Move Secure Adaptive Integrated Communications system, commonly called MOSAIC, is currently in the advanced technology demonstration stage of development. MOSAIC is intended to provide on-the-move net communications for the mobile, geographically dispersed battlefield. "Its wireless communications architecture will support multimedia applications; quality of service for mobile/multi-hop networks; adaptive and ad hoc mobility protocols; bandwidth management; and horizontal/vertical handoff in a mobile wireless environment" (U.S. Army, 2002a). These networks differ from civilian and most other networks in being ad hoc, since there can be no fixed hubs on a moving battlefield. Such systems would be very useful after an incident if there is a loss of civilian communications.

However, Objective Force C4ISR systems will need to be adapted for the different mission and different challenges of HLS. One difference between the Objective Force requirement and the future HLS C4ISR system is that unlike the former, which is designed to operate where no communications are available, the HLS C4ISR system may have the option of using an existing communications network—the nation's public switched network.

The public switched network may, however, be degraded following a major physical or cyber terrorist attack, so the future system should consider the expeditionary characteristics inherent in Objective Force concepts. Local connectivity might be gained in such a system through applications like the Joint Tactical Radio System and, perhaps, local, mobile laser communications networks, or transportable microwave networks, which would provide the bandwidth to share data and gain a common operational picture. The Army's WIN-T program, in development, can provide a seamless C2 grid where the local C2 infrastructure has been disabled.

The Army, working in conjunction with NORTHCOM, can provide the model for the national emergency response network. This model can set the standards for local and state C2 architectures, so that the DHS can seamlessly distribute critical information across the nation and to the agencies that need specific essential information to respond to threats and events. In the interim, the Army should investigate deployable communications packages equipped with universal multiplexer capability to facilitate C2 across the vast, and disparate, array of agencies that will respond to incidents and events.

Another promising development that the Army S&T community should address for the emerging HLS C2 system is Joint Blue Force Tracking (CJCS, 1999). The Blue Force Tracking architecture is designed to provide tracking, tagging, and locating of friendly troops and assets; logistics and asset management; and situational awareness. The Global Positioning System (GPS)-based concept can allow operational commanders to view the position of friendly forces in real time. Blue Force Tracking is being developed for U.S. forces engaged in expeditionary operations, but it could also be advantageous to know the location of local, state, and federal "forces" and key assets, including the Army, in real time, particularly when contending with the complex environment following a catastrophic event involving WMD.

The Army has explored many of the technologies necessary for an effective, end-to-end national emergency response C2 system. Applications of the S&T program essential to the Objective Force may provide a framework for such a system. If the system eventually adopted for the nation exploits and is compatible with Objective Force technologies, it can be beneficial to the Army. However, just as the Objective Force may have to operate with allies with various levels of modernization, the Army in discharging its HLS mission must address C2 compatibility with civilian responders. Table 4-1 highlights key S&T requirements for HLS C2.

Conclusion 4-1. A new national emergency response command, control, and communications system for homeland security must be developed and fielded to meet the demands of the emerging threats, particularly to integrate the response to chemical, biological, high explosive, radiological, and nuclear weapons. This system must be compatible with developments in the new Department of Homeland Security, the U.S. Northern Command, and state and local entities. Current Army science and technology thrusts and programs that are integral to the Objective Force can be adapted for the new national system.

Recommendation 4-1. To facilitate the development and fielding of an integrated command-and-control system for homeland security, the Army should initiate or continue research that permits the earliest possible fielding of deployable communications packages equipped with universal multiplexer capability to facilitate command and control across the vast, and disparate, array of agencies that will respond to incidents and events.

TABLE 4-1 Technologies for Command and Control

Functionality	Technology	Characteristics	Availability[a] (R, N, F)	Priority for Army S&T[b]	Multiuse[c] (H, O, C)
Command and control	Adaptive integrated communications	Multiplexer systems to integrate communications between multiple agencies	N	High	H, O, C
	Mobile local broadband networks	Mobile laser and/or microwave communications to pass imagery and communications	N, F	High	H, C
	Blue Force Tracking	System to determine the location of operational personnel and assets from multiple agencies	N, F	High	H, O, C
Planning	Decision support aids	Family of decision support aids such as those in the Agile Commander ATD to enhance real-time planning among multiple agencies for CM	N	High	H, O

NOTE: ATD, Advanced Technology Demonstration; CM, consequence management; TLR, technology readiness level.

[a]Availability: R, ready (TRL 8-9); N, near-term (TRL 4-7); F, far-term (TRL 1-3).

[b]Priority for Army S&T (investment): low, someone else has mission or technology is ready and available; medium, useful but of limited impact and some investment needed; high, very important, no one else working on it, considerable investment needed.

[c]Multiuse: H, Army homeland security; O, Objective Force; C, civilian (first responders and others).

Rapid Assessment of Physical Damage, Casualties, and Contamination

A necessary condition to conduct R and CM activities is an assessment of the situation. The Army and the DoD have introduced a program for the Family of Integrated Operational Pictures (FIOP). This program is designed to meet the needs of the warfighter. However, the concept could be extended to the HLS mission area, and the Army's experience with the Objective Force can help in doing so. Key elements for the development and fielding of an HLS common operational picture are the development and fielding of a family of both wide-area and focused sensors; the networking of these sensors for situational assessment; the fusion of sensor data; and adapting models that predict physical damage, contamination, and casualties based on real-time reports and sensor information. The situational awareness needed for HLS is closely related to the network-centric concepts inherent in the Objective Force; however, building such awareness is a complex problem because the operational picture must be shared by multiple agencies operating with mixed levels of systems and technologies.

A number of sensors exist that can assist with the real-time situational assessment. Overhead imagery from satellites and high-endurance unmanned aerial vehicles (UAVs) can build an optical and infrared picture of the physical damage. They can also use measurement and signal intelligence to determine WMD contamination. These assets provide a wide-area view of the "battle area." However, focused views of the affected area are needed. The family of tactical UAVs being fielded for the Objective Force can provide focused views of the HLS situation and be maneuvered to meet real-time needs of the on-scene commander. Chemical, biological, and radiological (CBR) surface sensors can be implanted throughout the affected area to fill in the picture. Robotic land vehicles can be used to implant and locate a family of surface sensors to characterize the damage. Finally, as the needs become more focused, sensors that can look into structures and detect casualties in rubble will need to be developed and fielded to complete the picture. Like the concepts and technology that underwrite the Objective Force, a common operational picture tailored to the demands of a specific contingency, integrated from wide-area sensors, filled in with tactically deployed air and land sensors, and augmented by specially designed and placed local sensors can help support the HLS mission.

The current state of sensors to characterize the effects and extent of CBR weapons varies. In Chapter 2, the committee describes the difficulties of detecting CBRNE weapons before they are employed. The post-attack assessment problem is easier technologically. However, it will be necessary to build the operational picture by networking multiple sensors and fusing the inputs into a common picture. For chemical weapons, local sensors are being fielded today, but there is still a need to improve the ability to characterize the attacks over a wide area. As we saw from the anthrax attacks in late 2001, a meticulous process of testing is necessary to identify the biological agent and to determine the extent of contamination. Nuclear and radiological detectors are the most highly developed sensors and can now be used to determine the extent of radiation.

Multiple sensor reports and images do not, by themselves, build the situational awareness and operational picture needed to conduct effective operations. The sensor pictures and reports need to be analyzed and depicted on a common grid and shared with the R and CM forces digitally. Fusion techniques are under development for the Objective Force, but here again the fusion technology for the HLS mission will need to be adapted to a related, but different, set of requirements. If such an information fusion capability is developed, it can also be used for warfighting in scenarios where WMD is threatened or actually used.

Finally, a family of models that can predict physical damage, contamination, and casualties can play an important role in the HLS mission. CBR contamination models today show the effects of known weapons. For example, the Army Risk Assessment Model system provides specific capability to examine the fate and transport of toxic materials in the environment and the implications for ecosystems and human health. The Anti-Terrorist (AT) Planner Software, developed by the U.S. Army Corps of Engineers (USACE) Research and Development Center (in conjunction with the Defense Threat Reduction Agency (DTRA) and the Technical Support Working Group), provides a flexible tool for examining the vulnerability of facilities to a variety of blast threats and the expected value of alternative approaches to enhance protection. The AT Planner is a good example of a technology whose use is currently restricted to the defense community (or contractors that serve the defense community) that could be of considerable use to the engineering community serving industry. However, the capabilities of these models need to be extended to predict contamination based on a limited set of reports and sensors readings. DTRA has a number of contamination models, and DARPA is also integrating models that can address this problem. These models are based on computational fluid dynamics approaches and their incorporation into simplified models that can be used to predict the movement of contaminants through the atmosphere, a city, inside buildings, and in tunnels and subway systems. Examples of such codes include the Hazard Prediction and Assessment Capability code (a dispersion code developed by DTRA that has been incorporated into its Integrated Munitions Effectiveness Assessment program), the Vapor, Liquid, and Solid Tracking (VLSTRACK) program, and the Dispersion and Diffusion Puff Calculator (D2PC).

As reported in *Making the Nation Safer: The Role of Science and Technology in Countering Terrorism,* work on this type of tool is proceeding, but results of the several models are often in disagreement. The report says, "Further R&D is needed to resolve these anomalies or develop more dependable alternatives" (NRC, 2002). The new technical challenge will be to link contamination models to real-time sensor reports and images to provide for timely attack assessment.

The civil engineering community possesses detailed structural drawings and models for civilian buildings and for facilities important to the nation's infrastructure. However, these are not readily obtainable in all localities and regions, nor can they be accessed in centralized databases. The application of these structural models along with progressive collapse technology can be used to forecast building failures and damage from terrorist attacks. The challenge will be to link these existing models to existing and emerging sensors that monitor structural health and to adapt them to the specific needs of the Army and the HLS community.

The Army should participate in and encourage the establishment of centralized databases that include structural drawings and models for high profile and critical infrastructure buildings and facilities. The databases would be used for assessing damage and casualty states in the event of terrorist attacks. The application of these structural models could forecast building failures such as occurred at the World Trade Center. Table 4-2 describes technologies for event assessment.

Conclusion 4-2. Rapid assessment of the effects of natural disasters and attacks using chemical, biological, high explosive, radiological, and nuclear weapons is essential to mitigate the damage, save lives, and restore order. To some degree, the process for event assessment is similar to that used by the Objective Force in building a common operational picture; however, different sensors and analytical processes will be used.

Recommendation 4-2. The Army should conduct research on processes and systems to facilitate the event assessment process. It should support the high-priority research such as sensor networking and fusion to merge reports from disparate sensors into a common picture.

Force Protection

The forces employed for large-scale R and CM activities need to be protected for sustained operations. Individual protection suits and inoculations are necessary to sustain operations in these conditions. The Army, through its Soldier and Biological Chemical Command (SBCCOM), continues to lead in the development of individual and collective protection technologies. The fielding of the Joint Service Lightweight Integrated Suit and the Joint Service Protective Mask over the next few years will provide some needed improvements in individual protection at a lower maintenance cost while relieving the physiological burdens of heat stress and breathing resistance. Current SBCCOM research on materials for facepieces and lenses, advanced filters, and service-life indicators to improve masks will aid the Army and the civilian community and should be aggressively continued.[1] Similarly, the research into protective clothing enhancements in-

[1]Anna Johnson-Winegar, Deputy Assistant to the Secretary of Defense (Chemical and Biological Defense), briefing to the American Association for Engineering Education Forum, Alexandria, Va., on February 25, 2002.

TABLE 4-2 Technologies for Event Assessment

Technology	Characteristics	Availability[a] (R, N, F)	Priority for Army S&T[b]	Multiuse[c] (H, O, C)
Family of interoperable operational pictures	Integrated situational awareness displays that can be shared by operational planners and implementers	N, F	High	H, O, C
Sensor development	Continued development of point and wide-area sensors to characterize chemical, biological, and radiological contamination following an attack	R, N, F	Low	H, O, C
	Development and fielding of sensors to determine the state of damage to buildings and to locate casualties in structures	R, N	Low	H, C
Robotics	Land mobile robotics that can breach obstacles to implant sensors that will characterize damage in a contaminated area	R, N	High	H, O, C
Sensor networking and fusion	Integration of multiple sensors into a common picture	N, F	High	H, O, C
Real-time modeling	Enhancement of damage and contamination models to provide attack assessments based on the reports of fused sensor data	N, F	High	H, O, C

NOTE: TRL, technology readiness level.

[a]Availability: R, ready (TRL 8-9); N, near-term (TRL 4-7); F, far-term (TRL 1-3).

[b]Priority for Army S&T (investment): low, someone else has mission or technology is ready and available; medium, useful but of limited impact and some investment needed; high, very important, no one else working on it, considerable investment needed.

[c]Multiuse: H, Army homeland security; O, Objective Force; C, civilian (first responders and others).

tended to reduce physiological stress, increase protection, and improve the logistics burden should maintain the priority given it by the Army. The direction of this research is to develop a family of selectively permeable membranes, reactive self-detoxifying materials, and electro-spun materials and to employ nanotechnology in this development effort.[2] Another promising concept for individual protection is the breast-pocket hood, which will provide survivors and first responders with crucial temporary protection from chemical and biological contamination.[3] Improvements in individual protection will assist the Army, the first responders, and other personnel who risk exposure following a terrorist event.

Mobile collective-protection facilities are necessary for long-term R and CM activities. The Army is currently developing a family of deployable collective-protection shelters that can be used by forces performing CM tasks, local and state authorities and their supporting workforce, and victims of the event (U.S. Army, 2002b, 2002c). Some of the collective-protection shelters are independent facilities that can be rapidly assembled; others are liners for existing buildings. The research that is under way in individual and collective protection is important both to the Objective Force and to the HLS mission.

The primary responsibility for the development of vaccines and medical countermeasures to protect against biological agents rests outside the Army, in the Department of Health and Human Services and the Centers for Disease Control. However, the expertise available in Army laboratories is essential to progress in this area, with the U.S. Army Medical Research Institute of Infectious Diseases in particular being a unique source of expertise and continued research. Table 4-3 describes technologies appropriate for force protection.

Conclusion 4-3. An aggressive, continuing science and technology program across the spectrum of technologies needed for individual and collective protection is necessary for the Army and civilian emergency responders.

Recommendation 4-3. The Army's research and development across the spectrum of technologies needed for individual and collective protection from the effects of weapons of mass destruction for the Army and civilian emergency responders should be continued.

Treatment of Mass Casualties

It is likely that mass casualties will result from the use of WMD and high explosive incidents. A mass casualty incident is one in which there are not enough

---

[2]Ibid.

[3]Corey M. Grove, Edgewood Chemical Biological Center, briefing to the committee on May 16, 2002.

TABLE 4-3 Technologies for Force Protection

Functionality	Technology	Characteristics	Availability[a] (R, N, F)	Priority for Army S&T[b]	Multiuse[c] (H, O, C)
Individual protection	Protective masks	Development of filters and service-life indicators for masks	R, N	High	H, O, C
	Suits	Development of semipermeable membranes and self-detoxifying material for protective suits	N	High	H, O, C
	Vaccines	Vaccine development for protection against biological agents	N, F	High	H, O, C
Collective protection	Mobile collective shelters	Enhancements to the family of collective shelters under development for the Objective Force	R, N	Low	H, O, C

NOTE: TRL, technology readiness level.

[a]Availability: R, ready (TRL 8-9); N, near-term (TRL 4-7); F, far-term (TRL 1-3).

[b]Priority for Army S&T (investment): low, someone else has mission or technology is ready and available; medium, useful but of limited impact and some investment needed; high, very important, no one else working on it, considerable investment needed.

[c]Multiuse: H, Army homeland security; O, Objective Force; C, civilian (first responders and others).

resources for casualty management. In the most likely scenarios, civilian emergency medical teams (EMT) and their field and individual equipment will be the first responders on the scene. Their first task will be to perform triage.[4]

Under normal circumstances, medics carrying out on-site triage have four responsibilities: (1) initiate the triage system and tag patients according to the severity of their injuries or illness, (2) report progress, intervention, and needs to the medical commander, (3) treat only immediate threats to life, i.e., blocked airways and severe arterial bleeding, and (4) move patients by priority to the casualty collection point.

In a mass casualty event, the triage[5] effort takes on an entirely different meaning, closely resembling rules of engagement in wartime or low-intensity conflicts. The approach will shift from the peacetime emphasis of optimized care for the individual to optimized care for the masses. The tasks of the EMT units will be to perform (1) initial high-level identification of life-threatening injuries and causes, stabilizing them whenever possible and appropriate, (2) assessment of on-going hazard and risk and or protection of responder personnel, (3) assessment of requirements for support infrastructure (facilities, communications, transportation, security), (4) medical triage, and (5) immediate medical response to the WMD event. Immediate medical response at the treatment center (civilian) relies on effective triage tagging. Continuing improvements in the techniques for triage and initial access (e.g., to patients trapped within confined structures), treatment, and distributed, secure communication will be necessary.

Where the cause of injury is suspected to involve chemical agents, toxins, or toxic industrial chemicals, the responders must be able to identify and evaluate whether the chemical is corrosive, ignitable, toxic, or reactive; subsequent actions and treatment by the medical responders will key from these observations. Methods for the field assessment of a biological hazard are also employed at this phase of the operation. Identification and containment of the agent after early presumptive diagnosis and identification of the threat will be very important because chemical and biological agents are indiscriminate and may be disseminated over large areas. The patient population will be diverse in age, gender, race, cultural preferences, and basal health. Further, the effects of biological agents can be particularly insidious in that they can be delayed, with the onset occurring and potentially contributing to distribution, even after the person has been transported to a safe area.

Communication of the identity and assessment of chemical and biological

---

[4]Triage is the sorting of patients by the severity of injury or illness so that resources can be more efficiently utilized to do the most good for the most people. Triage is conducted repeatedly: during the initial encounter with the civilian emergency medical teams, when the patient is stabilized, decontaminated, and moved to the casualty collection point.

[5]The Army's well-developed and validated approaches for triage could be adapted for civilian mass casualty emergencies.

agents and an estimate of ongoing risk to other components of the overall response teams will be critical for protection of the responders and for development of a perimeter of quarantine and its maintenance and eventual expansion. Timely and accurate information is essential to communicate instructions and guidelines to the public and to obtain its cooperation. A secure, independent communications system, vertical and horizontal integration of data, and decontamination of the patients and their tracking, along with tracking of physical evidence and clothing, over the event time line will be critically important.

Additionally, decision aids to determine dynamic disaster response and evacuation and quarantine policies tailored to the tactical situation will be needed. Medical personnel treating victims of WMD will probably require support from remote experts to identify the chemical or biological agent used in WMD events, including on-demand linkage to medical and scientific information systems, experts, and laboratories. Further, the sharing of acquired insight (agents, medical implications and treatments, exposure/decontamination data, patient and patient property tracking, etc.) will require a chain of custody and will probably be important for building an overall picture of the event theater.

While it is essential that the military capability be able to interface with civilian HLS capabilities as needed, some aspects of the military capability may not perfectly match HLS applications. For example, material designed to meet warfighter requirements may not be suitable for civilian use because of material or training constraints. OSHA must certify personnel protection equipment for civilian use, and medical products for distribution to civilians must be fully licensed by the Food and Drug Administration or used with individual informed consent. Military medical defense products for CBRNE assault assume a healthy adult population, but civilian populations exposed to terrorist assault will vary in health and age. Some defense vaccines, pretreatments, and post-event treatments may confound other medical treatments and cultural/religious preferences. Moreover, pre-exposure immunization of large populations against biological agents may not be warranted. Finally, full voluntary compliance cannot be guaranteed for a large civilian population. Application of Army S&T to HLS medical needs will have to address these issues.[6] Table 4-4 describes technologies for medical response.

> **Conclusion 4-4.** The new challenges for recovery and consequence management include triage, tracking, and treatment of mass casualties following an event involving weapons of mass destruction. The scale of such an event

[6]Anna Johnson-Winegar, Deputy Assistant to the Secretary of Defense (Chemical and Biological Defense), briefing to the American Association for Engineering Education Forum, Alexandria, Va., on February 25, 2002.

and the need to conduct an orderly treatment process in the presence of chemical, biological, radiological, or nuclear contamination is daunting. In all likelihood, the nation's military, including the Army, will be called on to play a significant role in this activity.

Recommendation 4-4a. The Army should expand its research in the area of triage, tracking, and treatment of mass casualties.

Recommendation 4-4b. The Army should ensure development of individual triage assessment for mass casualties from events involving weapons of mass destruction.

Recommendation 4-4c. The Army should ensure the development of a process to leverage information technology to effectively conduct mass casualty triage, tracking, and treatment following such an event. The process development should incorporate remote decision support systems that can be integrated with civilian systems, and a tracking system.

Containment and Decontamination of the Effects of Weapons of Mass Destruction

There is not much experience in wide-area decontamination of the effects of CBRN weapons. Even if the levels of contamination can be assessed, there are few tools or techniques available for such broad decontamination. One has only to look at the difficulty of sanitizing the facilities contaminated with the anthrax virus in late 2001 to be reminded of this. Chemical and radiological contamination present equally daunting challenges.

Decontamination will probably be accomplished in stages, and it is likely that the Army will be involved in early remediation of the effects in WMD events. Decontamination will be a time-critical and stressful task. First, the extent and toxicity of contamination must be determined. It is also likely that the cleanup tasks will be accompanied by substantial physical damage and the need to provide care for mass casualties. Complicating the difficulty of the decontamination process is the fact that standards for cleanup and decontamination have not been developed, although models do exist from civilian cleanup following toxic waste accidents. A structured process based on a real-time attack assessment will be needed to conduct decontamination and cleanup operations.[7] For chemical and biological events, a suite of technologies is available:

[7]John F. Weimaster, Director, Research and Technology Directorate, Edgewood Chemical Biological Center, U.S. Army Soldier and Biological Chemical Command, briefing to the committee on July 18, 2002.

TABLE 4-4 Technologies for Medical Response

Functionality	Technology	Characteristics	Availability[a] (R, N, F)	Priority for Army S&T[b]	Multiuse[c] (H, O, C)
Individual triage assessment and tracking	Chemical, biological, and radiological triage assessment cards	C4ISR integration of data, decontamination of the patients and material, tracking of the patients, physical evidence, clothing; chain of custody	R, N	High	H, O, C
Triage decision support integrated with civilian systems	C4ISR; on-demand access to expert's network, scenario modeling/procedures	Remote expert support for the on-site medical personnel; on-demand linkage to medical and scientific information systems, experts, and laboratories. Vertical/horizontal sharing of insight (agents, medical implications and treatments, exposure/decontamination)	R, N	High	H, O, C
Medical support systems	Field-deployable diagnostic, life-support, and emergency surgical systems	Systems that can be easily and rapidly deployed; that are resistant to vibration, low environmental quality and electromagnetic interference; and that can be operated efficiently in the presence of the assaulting weapon (chemical, biological, radiological residuals)	R, N, F	High	H, O, C
Environmental monitoring and threat assessment tools	Field-deployable rapid assay devices; dynamic meteorologic models of CBRN threats	First responder assessment of agents and risks for staff and patients; assessment of ongoing environmental risks	R, N	High	H, O, C

Toxicological models for exposure to CBRNE agents	Scenario development software based on physiologic and biochemical response to agents	Field support for identification of assault agents and probably course of development	R, N	High	H, O
Conventional therapeutics	Hemorrhage, neurological, and respiration stabilizing devices and technologies	Long shelf-life, rapid acting agents	R, N	High	H, O, C
Individual countermeasures	Vaccines and immunologic factors (including therapeutic application), counteragents for chemical, biological, and radiological exposure	Long shelf-life, rapid acting agents	R, N, F	High	H, O
MCI training platforms	Distributed learning platforms with AI and decision-assisting tools for CBRNE		R, N, F	High	H, O

NOTE: AI, artificial intelligence; C4ISR, command, control, communications, computers, intelligence, surveillance, and reconnaissance; CBRN, chemical, biological, radiological, and nuclear; CBRNE, chemical, biological, radiological, nuclear, and high explosive; MCI, mass casualty incident; TRL, technology readiness level.

[a]Availability: R, ready (TRL 8-9); N, near-term (TRL 4-7); F, far-term (TRL 1-3).

[b]Priority for Army S&T (investment): low, someone else has mission or technology is ready and available; medium, useful but of limited impact and some investment needed; high, very important, no one else working on it, considerable investment needed.

[c]Multiuse: H, Army homeland security; O, Objective Force; C, civilian (first responders and others).

- Gas-phase decontaminants such as chlorine dioxide and vapor-phase hydrogen peroxide,
- Solution chemistry—chlorine and hypochlorite formulations, oxidative systems like hydrogen peroxide, and
- Catalytic systems such as enzymes.[8]

There are examples of responses after radiological/nuclear events, but they are limited. The cleanup following the B-52 accident at Palomares, Spain, stands out as the primary practical example of radiation cleanup by the United States. The nuclear decontamination process at Chernobyl may also provide some useful lessons learned. The common denominator in radiological decontamination is that the particles must be contained, encapsulated, and physically removed from the area at some point.

Considerable research, process development, training, and planning will be necessary to successfully conduct decontamination following a CBRNE event. The Army, and perhaps the Department of Energy, will be at the forefront of the research necessary to build this capability. Table 4-5 describes technologies for remediation and decontamination.

Conclusion 4-5. The processes for decontamination following chemical, biological, radiological, nuclear, or even large explosive events need to be expanded. Rapid remediation of the areas involved in such an event will be necessary to limit casualties and to restore critical services. Expanded Army science and technology can contribute significantly to process development and to finding decontamination materials to assist the activity.

Recommendation 4-5a. Army science and technology should concentrate on the further development of a process to plan and implement remediation and decontamination for chemical, biological, radiological, and nuclear events. This process must be capable of being conducted in real time based on limited information.

Recommendation 4-5b. Army science and technology should concentrate on the further development of decontamination solutions for chemical, biological, nuclear, or even large explosive events.

SUMMARY

The challenges of R and CM posed by a massive domestic terrorist event present the Army with new requirements for S&T. There is a high degree of

[8]Ibid.

TABLE 4-5 Technologies for Remediation and Decontamination

Technology	Characteristics	Availability[a] (R, N, F)	Priority for Army S&T[b]	Multiuse[c] (H, O, C)
Decontamination process development	Development of a process to plan and implement remediation and decontamination for chemical, biological, radiological, and nuclear events	N	High	H, C
Decontamination solutions	Further development and assessment of solutions to clean up chemical and biological contamination	R, N, F	High	H, C

NOTE: TRL, technology readiness level.

[a]Availability: R, ready (TRL 8-9); N, near-term (TRL 4-7); F, far-term (TRL 1-3).

[b]Priority for Army S&T (investment): low, someone else has mission or technology is ready and available; medium, useful but of limited impact and some investment needed; high, very important, no one else working on it, considerable investment needed.

[c]Multiuse: H, Army homeland security; O, Objective Force; C, civilian (first responders and others).

overlap with the research and development already under way for the Objective Force; however, R and CM for HLS will require adaptations of the current thrusts and, in some cases, new S&T. In some areas, other government agencies and the private sector can be expected to conduct the S&T, but the Army will have to monitor developments and then adapt the results to its specific needs.

REFERENCES

CJCS (Chairman of the Joint Chiefs of Staff). 1999. CJCSI 8910.01, Blue Force Tracking Collection and Dissemination Policy, December 15. Washington, D.C.: Office of the Chairman of the Joint Chiefs of Staff Public Affairs.

NRC (National Research Council). 2002. Making the Nation Safer: The Role of Science and Technology in Countering Terrorism. Washington, D.C.: National Academies Press.

U.S. Army. 2002a. United States Army Weapon Systems 2002. Washington, D.C.: Government Printing Office.

U.S. Army. 2002b. M20A1 Simplified Collective Protection Equipment (SCPE). Available online at <http://www.sbccom.apgea.army.mil/products/m20a1.htm>. Accessed on October 15, 2002.

U.S. Army. 2002c. M28 Simplified Collective Protection Equipment (CPE). Available online at <http://www.sbccom.apgea.army.mil/products/m28.htm>. Accessed on October 15, 2002.

White House. 1998. The Clinton Administration's Policy on Critical Infrastructure Protection: Presidential Decision Directive 63, May 22. Available online at <http://www.cybercrime.gov/white_pr.htm>. Accessed on October 3, 2002.

5

Attribution and Retaliation Technologies

INTRODUCTION

In general, attribution is assigning a cause or source to an act or event. In the context of this report, it is the identification of individuals or organizations that are responsible for direct or indirect acts of terrorism and sabotage directed against the United States, its territories, and vital national interests, and those that support them. Attribution is dominated by operations that identify those responsible; their tactics, techniques, and procedures; their equipment, materiel, and logistics; and their operational locations.

Retaliation is action taken in return for an injury or offense and future deterrence. For this report, it is defined as those operations that are focused on capturing, killing, and eliminating those individuals, organizations, their supporters, and their operational ability to conduct acts of terrorism and sabotage directed against the United States, its territories, and vital national interests. Additionally, these operations aim to create effects and demonstrate consequences that will deter other groups that might plan such attacks and to bring any such perpetrators or their supporting agencies, organizations, or foreign governments to justice.

OPERATIONAL AREA AND THE ARMY ROLE

The Army's role in homeland security (HLS), antiterrorism (AT), and counterterrorism (CT) is addressed specifically in Chapter 1. The tasks include the following:

- Force protection of soldiers, families, and installations;
- Operations in support of the lead federal agency or state in case of a large-scale conventional or weapons of mass destruction (WMD) attack; and
- Operations in support of Joint Military Operations.

The Army's particular role in attribution is very limited, both at home and in host nations. The intelligence community, whether the Federal Bureau of Investigation or the Central Intelligence Agency or any of the other less well known agencies, will make the attribution. The Army's role is primarily that of support, either providing perimeter security and crime scene protection or providing analytical support from one of the Army's premier technical labs.

In contrast, the Army's role in retaliation runs the gamut from simple military/law enforcement coordination in the United States, when appropriate, to full-blown remote combat operations overseas, where the Army may be assigned primary responsibility for ground retaliation. Since this role is a primary one for the Army, the committee believes there are some enabling technologies that should receive very high priority and deserve S&T investment.

TECHNOLOGY FOCUS AREAS

Because the potential range of response is so broad, the committee feels it would be most useful to focus on three limited areas that present very difficult technical challenges and where S&T can act as a force multiplier:

- Remote operations in an urban environment, with focus on mobility and survivability,
- Situational awareness in urban environments, and
- Terrorist surveillance in difficult environments, both urban and rugged[1] terrain.

Remote Operations in an Urban Environment

As mentioned above, the committee believes that technology can be extremely useful to the Army in urban operations. In particular, technology can enhance mobility, survivability, and precision fire support.

- *Mobility.* Moving quickly in a crowded city swarming with civilians and hiding some terrorist cells is an extremely complicated task. This

[1]By "rugged" the committee means dense foliage or hilly terrain where it is difficult to use overhead assets or organic platform sensors to find terrorist cells.

problem was clearly demonstrated in Somalia. The Army must be able to move personnel quickly, through or over busy streets, on a safe, survivable platform. There is a need for small, armor-plated, light transport vehicles, ground and helicopter, to move forces as needed in this environment. Additionally, the capability is needed to clear obstacles in the streets and alleyways.

- *Survivability.* There are several aspects to the survivability problem. One key aspect is signature reduction of our forces across the spectrum—radio frequency (RF), electro-optical, infrared, radar, acoustic, etc. Success here could have a major impact on survivability. Additionally, enhanced armor protection is a must. Investment in very light but immensely strong armor can make a big difference and ought to be funded accordingly. This is also of critical importance in the Objective Force Warrior program (U.S. Army, 2002).
- *Fire support.* Fire support plays a critical role in all combat operations. Most current fire-support systems were not developed specifically for urban warfare, where precision and lethality (or nonlethality) are significant factors in the outcome of an operation. Even relatively small errors can be devastating in terms of collateral damage or innocent civilians killed. Continued development of precision munitions and adaptations of all fire-support weapons with both Global Positioning System (GPS) and GPS-type tracking is a must. Additionally, the issue of lethality must be addressed.

Often, traditional means of fire support can be used effectively in urban combat. However, even with more precision, fire support from systems such as AC 130 gunships, AH 64 attack helicopters, and artillery may not provide the immediate dedicated and more delicate support required by troops on the ground. In urban combat, the right tool may be a tack hammer, not a sledgehammer. The Army S&T program should explore concepts such as unmanned aerial vehicles (UAVs) or unmanned ground vehicles (or both) that can loiter one block away and be called forward by the ground commander when needed. They can provide not only lethal but also nonlethal support in the form of concussion grenades, incapacitating agents, or psychological operations products.

Conclusion 5-1. Lack of mobility in an urban environment is a critical disadvantage that can result in survivability challenges.

Recommendation 5-1. The Army should continue and enhance current research and development to focus on mobility operations in the urban environment, to include exploration of small, mobile armored carriers for use in urban environments and mini-breachers to clear streets and alleyways.

Conclusion 5-2. Precision and lethality of weapons are critical issues the Army should address to improve fire support for operations in urban environments.

Recommendation 5-2. The Army should modify current systems or develop new systems, along with appropriate munitions, that are specifically designed for extremely precise fire support in urban environments.

Situational Awareness in Urban Environments

The current system for gaining situational awareness in an urban environment is inadequate. This is due to the extremely complex RF propagation environment in such a setting, coupled with the high-resolution accuracy needed to track a soldier in a specific room or building. A comprehensive situational awareness system is needed. Building on the current Land Warrior system (U.S. Army, 2002), such a system would link the individual soldier to on-the-body, local, and remote sensor systems and information databases.

Conclusion 5-3. Several capabilities and technologies being developed by the Army would be extremely useful for the civilian first responder, for example the situational awareness Blue Force Tracking and health monitoring system.

Recommendation 5-3. The Army should make technologies such as the situational awareness Blue Force Tracking program and the health monitoring system available to the Department of Homeland Security, which will consider whether or not they can be adapted for civilian use.

Elements of such a situational awareness system need to include:

- *High-resolution blue force (friendly) tracking.*[2] Current systems have inadequate resolution and are unable to exactly locate the individual soldier inside a building or a room due to the complex RF environment and lack of resolution of the GPS system. What is needed is a more complex system relying on GPS, a local RF system, and an accurate dead-reckoning system.
- *Surveillance sensors.*[3,4,5] There is no sufficiently lightweight, robust, multi-sensor, low-power, low-bandwidth sensor system. Such a system should provide information (and video imagery) to soldiers both as trig-

[2]For instance, the U.S. Army Communications-Electronics Command's (CECOM's) Counter Terrorism Blue Force Situation Awareness Protection Suite, briefed to the committee by Raymond Filler, CECOM Research and Development Engineering Center, May 16, 2002.

gered by enemy activity and as requested by the soldier. It should include several sensor phenomenologies, to wit: infrared, low-light-level visible, acoustic, seismic, chemical, and biological. Additionally, it should include advanced sensor fusion algorithms that provide composite automatic target recognition and identification systems, and, to reduce the workload on an operator, it should include alternatives to current video systems that rely on pan-tilt-zoom capability (such as panoramic systems). Lastly, it should include sensors that can monitor tunnels and locate booby-traps.

- *Information databases.* Systems should provide access to archival information about the urban environment, including: (1) building structures, (2) street maps, (3) the transportation network, (4) weather data, and (5) blueprints for individual major buildings.
- *Red force (enemy) information processing and fusion.* The intelligence officer can be and will be quickly overloaded with an abundance of red force spot reports. The system needs tracking/deconfliction algorithms, which would allow for the detection of intruders, tracking, recognition (uniform and face recognition techniques), and time/space correlation of intrusion events to determine size and activity.
- *Red force location/tracking.*[6] The current capability is very limited. Such advances as cellular phone intercept and tracking and through-wall sensing (RF and acoustic) would help significantly if they can be made small, lightweight, inexpensive, and effective.
- *Health monitoring system.* There is a need to monitor the key parameters of individual soldiers' health, such as body core temperature, hydration level, heart rate, biological and chemical exposure levels, and wound location and severity, utilizing a system like the one being developed by the Army Institute of Environmental Medicine (U.S. Army, undated). Additionally, predictive models of human stress failure points as a function of measured parameters would be useful for the commander.

---

[3]For example, the U.S. Army Night Vision and Electronic Sensors Directorate's technology programs for counterterrorism (CT Echelon Surveillance and Reconnaissance, Multi-Function Remote Unattended Ground Sensors (CECOM I2WD), Remote Observation and Confirming Sensor, Cave/Urban Assault Kit, Advanced Search and Rescue Technologies, Cave/Urban Assault ACTD), briefed to the committee by A. Fenner Milton, Night Vision and Electronic Sensors Directorate, May 15, 2002.

[4]For example, the U.S. Army Research Laboratory's LIBS Sensor for Field Detection of All Hazardous Materials, briefed to the committee by Roy Walters, Director of R&D, U.S. Army Research Laboratory, May 16, 2002.

[5]For example, see Networked Sensors for the Objective Force ATD in U.S. Army, 2002.

[6]For example, the U.S. Army CECOM sense-through-the-wall technology, briefed to the committee by Robert Foresta, Branch Chief for SIGINT Payload and Integration Division, Intelligence Collection Branch, CECOM, May 15, 2002.

Conclusion 5-4. A very sophisticated situational awareness system, with highly accurate Blue Force Tracking in an urban environment, although difficult to construct due to complex radio frequency characteristics and the degree of accuracy required, will provide the soldier and civilian emergency responders a very powerful tool in the war against terrorism.

Recommendation 5-4. The Army should continue to develop a robust soldier situational awareness system begun in Land Warrior that provides a real-time, fused information system.

Terrorist Surveillance and Tracking (Rugged Terrain)

Locating and tracking small terrorist cells in a rural environment is a very difficult task, particularly when the terrorists attempt to blend into the environment. This is the detection issue addressed in Chapter 2: technically speaking, a very small signal against a large background. Several advanced technologies may help the war fighter locate terrorists in this environment:

- *Advanced unattended ground sensors (UGS).*[7] Remotely replaceable, power efficient, multisensor unattended ground systems will allow the war fighter to gather data and monitor critical locations, such as a crossroad, transportation junctions, critical building and gathering sites, etc. These systems must be covert, remotely enplaceable (perhaps robotically), preferably redeployable, and able to run in low-power mode until keyed by some event. They should have the ability to be integrated into, and cued by, a higher echelon information system that uses airborne and spaceborne assets, one element of the emerging network-centric warfare system.
- *Multipayload, multisensor UAV surveillance system.*[8] A (preferably) covert UAV system can be rapidly deployed for surveillance in areas of interest. The payload should be multiple sensors able to detect covered and concealed targets. Power, size, and weight issues are paramount. Sensor fusion algorithms are necessary. They should include foliage penetration (FOLPEN) systems. Such a system will be fairly challenging technically, given the weight, power, and size needed to provide effective

[7]For example, the U.S. Army Night Vision and Electronic Sensors Directorate's technology programs for counterterrorism (CT Echelon Surveillance and Reconnaissance, Multi-Function Remote Unattended Ground Sensors (CECOM I2WD), Remote Observation and Confirming Sensor, Cave/Urban Assault Kit, Advanced Search and Rescue Technologies, Cave/Urban Assault ACTD), briefed to the committee by A. Fenner Milton, Night Vision and Electronic Sensors Directorate, May 15, 2002.

[8]For example, see Networked Sensors for the Objective Force ATD in U.S. Army, 2002.

coverage over a large area. False-alarm reduction will be a significant challenge.

However, as discussed in Chapter 2, there may well be a physical limitation to detector capability. The committee suggests the following approach in this case: First, look at increasing sensor sensitivity. Where that does not provide sufficient gain, look at networking lower cost, distributed sensors to cover a broad area. Where that is not sufficient, look at fusing disparate sensors such as ground sensors and airborne sensors to increase sensitivity. Where that doesn't work, look at information fusion, i.e., combine results from different sources such as human intelligence reports, abnormal activity (heavy traffic or unfamiliar vehicles), overhead assets, and local sensors.

Conclusion 5-5. Terrorist cell tracking and surveillance in the urban environment and in rugged terrain are extremely difficult as they rely on a very small signal against a large background.

Recommendation 5-5. The Army should adopt a tiered approach to the problem of terrorist cell tracking and surveillance in the urban environment and in rugged terrain, first increasing sensor sensitivity, then networking and fusing sensors, and, finally, fusing information from disparate sources.

GENERAL FUNCTIONALITY, TECHNOLOGY, AND PRIORITY

In this section, the committee summarizes the general functionalities associated with the Army's role in attribution and retaliation and lists the technologies that could support their accomplishments seen by the committee (see Tables 5-1 and 5-2). The list is not meant to be all-inclusive. Moreover, the priorities are somewhat subjective, reflecting, as they do, the opinions of the committee. They reflect the committee's assessment of the importance of the specific task of accomplishing the Army's mission and the importance of the technology to accomplishing the task. Consequently, where the Army role is secondary, as it is in almost all aspects of attribution, the priorities are at best medium. Where the Army has primacy, as it does in many aspects of retaliation, and where the technologies may provide a leap-ahead capability in accomplishing the tasks, the priorities are high. As discussed in Chapter 1, chemical and biological investments are managed at the Office of the Secretary of Defense level. Similarly, some-far term technologies that are high risk but high payoff are nominally the province of the Defense Advanced Research Projects Agency. Where the technology would primarily support first responders, the committee believes the Department of Homeland Security should be the lead agency.

The tables provide a collection of technologies that could be used during the attribution and retaliation phases. The availability column reflects the general

TABLE 5-1 Technologies for Attribution

Functionality	Task	Technology	Availability[a] (R, N, F)	Priority for Army S&T[b]	Multiuse[c] (H, O, C)
Incident analysis	Biological analysis support	Enhanced biological analysis tools	F	Low (OSD)	H, C
	Chemical analysis support	Enhanced chemical analysis tools	N	Low (OSD)	H, C
	Explosive analysis support	Enhanced explosive analysis tools	N	Low (DHS)	H, C
	Cyber analysis support	Enhanced software analysis tools	N	Low (DHS)	H, O, C
Military/ law enforcement coordination	Database interoperability (domestic and host nation)	Software interface tools	N	Low (DHS)	H, O, C
	Communications interoperability	Hardware/software development	N	Low (DHS)	H, O, C
	Protect crime scene	Improved intrusion detection system	N	Low (DHS)	H, C

NOTE: OSD, Office of the Secretary of Defense; DHS, Department of Homeland Security; TRL, technology readiness level.

[a]Availability: R, ready (TRL 8-9); N, near-term (TRL 4-7); F, far-term (TRL 1-3).

[b]Priority for Army S&T (investment): low, someone else has mission or technology is ready and available; medium, useful but of limited impact and some investment needed; high, very important, no one else working on it, considerable investment needed.

[c]Multiuse: H, Army homeland security; O, Objective Force; C, civilian (first responders and others).

TABLE 5-2 Technologies for Retaliation

Functionality	Task	Technology	Availability[a] (R, N, F)	Priority for Army S&T[b]	Multiuse[c] (H, O, C)
Surveillance and tracking	Detect traffic/activity abnormality in urban and rural locations	Multisensor fusion	N	High	H, O
		Data mining techniques	N	High	H, O
		Inference algorithms	N	High	H, O
		Redeployable UGS	F	High	H, O
	Locate terror cells in areas of heavy foliage	3-D ultrasensitive lidar	N	High	O
		FOLPEN radar on UAV	F	Low (DARPA)	O
		Redeployable UGS	F	High	O
		Hyperspectral effluent detection	N	Medium	O
		Cell phone tracking	N	Low (NSA)	O
	Defeat covered and concealed targets in rural environment	3-D ultrasensitive lidar	N	High	O
		Multisensor fusion techniques	N	High	O
	Locate terror cells and personnel in buildings	Through-wall radar	F	Medium (DARPA)	H, O, C
		Ultrasensitive acoustics location systems	F	Medium (DARPA)	H, O, C
		Cell phone tracking	N	Low (NSA)	H, O, C
	Locate gunshots in urban environment	Ultrasensitive acoustics triangulation system	F	High	H, O, C
	Enhanced red force (enemy) location in urban environment	RF tags	N	Low (DARPA)	O
		Track deconfliction algorithms	F	High	O
		Personnel automatic target recognition algorithms	N	Medium (DARPA)	O

Situational awareness	Enhanced blue force (friendly) personnel location in urban environment	Fused GPS, RF, and dead-reckoning hardware and algorithms	N	High	H, O, C
Military/law enforcement coordination	Nonlethal apprehension tools	Incapacitating gas	N	Low	O, C
		Improved rubber bullets	N	Low	O, C
		Sticky foam	N	Low	O, C
	Identification	Rapid DNA analysis	N	Low	O, C
		Interpol/interagency database access	F	Low	O, C
		Rapid fingerprinting analysis	F	Low	O, C
		Face recognition algorithms	F	Low	O, C
	Interrogation support	Universal translator	F	Low (DARPA)	O, C
		Advanced lie-detection techniques	F	Low (DARPA/ DHS)	O, C
Remote operations	Mobility in urban environment	Exoskeleton for soldier platform	F	High	O, C
		Lightweight, highly survivable ground platform	F	Low (DARPA)	O, C
		Light, highly survivable, signature-suppressed troop-carrying helicopter	F	High	O, C
		Mobile, small-scale robotic breachers for clearing alleys, etc. in urban environment	N, F	High	O, C
	Signature reduction; lower all signatures	Low-signature RF, acoustic, EO, IR, radar	F	Low (DARPA)	H, O, C
		Reduced usage of signature-producing technologies	N	High	H, O
	Robots; improved CBRN detection and protection	Individual biological agent monitor and reporting system	F	Low (OSD)	H, O, C
	Lightweight armor protection	Advanced composites	F	High	H, O, C

Continues

TABLE 5-2 Continued

Functionality	Task	Technology	Availability[a] (R, N, F)	Priority for Army S&T[b]	Multiuse[c] (H, O, C)
	Enhanced vehicle mine protection	Advanced composites	F	High	H, O, C
	Advanced health and wound monitoring system	Integrated blood pressure, heart rate, body temperature, skin penetration sensors	N, F	High	H, O, C
	Rapid, automatic resupply	Automatic logistic resupply algorithm	N	Medium	O
		Robotic resupply system	F	Low (DARPA)	
	Munitions and delivery systems designed for urban combat	Nonlethal munitions to include acoustic systems	N, F	High	H, O, C
		PSYOP products	N	High	O
		UAVs and UGVs designed for urban fire support	N	High	H, O, C
Precision insertion and targeting	Improved warheads	Advanced propellants	N, F	High	O
		Improved warhead design	N, F	High	O
	Improved precision	GPS, RF, and remote laser designation systems	N	Medium	O

NOTE: UGS, unattended ground sensors; 3-D, three-dimensional; lidar, light detection and ranging; FOLPEN, foliage penetration; UAV, unmanned air vehicle; DARPA, Defense Advanced Research Projects Agency; NSA, National Security Agency; RF, radio frequency; GPS, Global Positioning System; EO, electro-optical; IR, infrared; CBRN, chemical, biological, radiological, and nuclear; PSYOP, psychological operations; UGV, unmanned ground vehicle; TRL, technology readiness level; DHS, Department of Homeland Security.

[a]Availability: R, ready (TRL 8-9); N, near-term (TRL 4-7); F, far-term (TRL 1-3).

[b]Priority for Army S&T (investment): low, someone else has mission or technology is ready and available; medium, useful but of limited impact and some investment needed; high, very important, no one else working on it, considerable investment needed.

[c]Multiuse: H, Army homeland security; O, Objective Force; C, civilian (first responders and others).

opinion of the committee, which is not meant to be a detailed evaluation. A more precise evaluation, including a risk assessment of the technology, would be the province of a follow-on study.

REFERENCES

U.S. Army. Undated. Warfighter Physiological Status Monitoring. Available online at <http://www.usariem.army.mil/wpsm/index.html>. Accessed January 13, 2003.

U.S. Army. 2002. United States Army Weapon Systems 2002. Washington, D.C: Government Printing Office.

6

Committee Observations

The U.S. Army is facing a challenge. Just as it launches a transformation toward the Objective Force, the centuries-old responsibilities for support to civil authorities have again been brought to the fore by the terrorist attacks of September 11. The committee found that these apparently diverse requirements are actually resulting in important convergences of technical and operational solutions. The requirements of homeland security (HLS) can for the most part be met through S&T work already set in motion for the Objective Force. The events of September 11 have stressed the Army's S&T planning and budgeting and are necessitating a reconsideration of the process by which the S&T Master Plan is being developed and a review of its contents. While many, if not most, of the Objective Force technologies are of direct application to the Army's recently reconfirmed homeland responsibilities, it will be necessary to modify or adapt specific technologies to serve a dual purpose. In addition, some new capabilities requiring modified acquisition strategies will be needed. The committee believes that if this process is accomplished thoughtfully and flexibly, there will be great opportunities for cost-effective procurements, economies of scale, and an ability to accomplish both missions successfully.

Throughout this report the committee has reached findings and conclusions and offered a series of recommendations on specific aspects of the HLS challenge for the Army. (All the chapter findings, conclusions, and recommendations are listed in numerical order in Chapter 7.) In this chapter, the committee summarizes its high-level integrated observations.

Defense of the homeland is the military's top priority; terrorism will increase the Army's efforts in support of civilian authorities (see Chapter 1). The committee reviewed the roles of the three Army components—the active Army, the Army National Guard, and the Army Reserve—in homeland emergencies. Various units of the Army are regularly used in natural disasters such as floods, fires, and hurricanes and tornadoes. The committee believes that terrorism will greatly enlarge the need for Army resources in the homeland.

The Army National Guard component will have a prominent role in homeland security; this growing role is not recognized in the annual development of the Army Science and Technology Master Plan (Chapter 1). The committee reviewed the restrictions on the federal portions of the Army under the Posse Comitatus Act and found that the National Guard, under state control, is the natural Army component to address terrorist attacks, at least initially. The committee also observed that the National Guard's technical needs for performing this role have not received uniformly high priority. It notes that certain specialized elements of the active and reserve Army are regularly employed in disasters in an other than law enforcement role, such as the engineering, medical, and logistics units. The magnitude of this effort will increase in the face of terrorism.

Many of the technologies recommended by the committee for use by the Army for HLS are also of high priority in the R&D plans for the Objective Force (Chapter 1). The committee believes that this overlap of technical needs should make it easier to develop R&D investment strategies in both areas. The committee also believes increased R&D in sensors; in communications, command, and control; and in the medical arena, all three of which are common to HLS and the Objective Force, will be helpful. While the details will likely differ, necessitating R&D to adapt from one area to the other, the substance will be the same.

There are striking similarities between the active Army working with allies and coalitions of allies and the HLS requirement for the Army to work with state and local civilian emergency responder organizations (Chapters 1, 4, and 5). Many of these challenges are technical; many are cultural. The committee focused on the technical but is concerned about the nontechnical issues that may operate to the detriment of close working relationships. This is especially true in communications, command, and control, where there are difficult organizational and operational challenges.

The committee observed that the technologies included under command, control, communications, computers, intelligence, surveillance, and reconnaissance (C4ISR) are of prime importance for HLS and for the Objective Force (Chapters 1, 4, and 5). Difficulties with interoperability have been encountered in both areas, sometimes with devastating results. The Army may have to replace interrupted civilian communications services on an emergency basis. This will require downward-compatible, plug-in capabilities.

The committee observed that rapid event assessment is essential in HLS in order to mitigate losses (Chapter 4). The responders first on the scene need a means of rapidly knowing what kinds of hazards are present. Technologies are needed to assess rapidly and accurately the nature of the threat, its extent and severity, and the changes of these variables with time. In the case of fire, the responders need to know instantly the rate of growth and the effect on safety within a structure.

The technologies for situational awareness for the Objective Force can be adapted for use by civilian site commanders at scenes of terrorism (Chapter 4). The committee believes the need to know where the first civilian emergency responders are, what they are doing, and where they are moving is the same as the military's need to know where their forces are on the battlefield. Civilian incident commanders need to know the location and movements of individual responders, such as firefighters inside buildings.

The committee finds that methods of sensing specific threats in chemical, biological, radiological, nuclear, and conventional explosive and incendiary weapons are not adequate to combat terrorism (Chapter 2). Packaged nuclear devices, explosives, and biologicals are particularly difficult to detect even when the detector is close to the package. Chemicals, because of their higher vapor pressure, are somewhat easier to detect. The committee believes that new technical approaches are needed and emphasizes smart networks of multifunctional detectors. Given the all-encompassing role of such detectors, the committee believes they are legitimate research topics for the Army even though some of the functions are in areas assigned elsewhere.

The committee has identified high-priority areas for R&D that could significantly reduce losses at Army facilities due to blast and impact (Chapter 3). R&D advances can minimize the chances for progressive collapse, improve structural connections, reduce dangerous debris from window and wall materials, and improve design practices for multihazard situations. The committee believes that a serious effort must be made to transfer new technologies in this area to civilian designers and contractors.

The Army should continue to give the highest priority to cybersecurity and to the use of best practices (Chapter 3). One disaster scenario envisioned by the committee involved terrorists operating over computer networks to shut down or alter targeted computer systems. The likely effects in the committee's scenario were interruption of DoD command-and-control systems; loss of power across the national electricity grid; denial of service over the public switched network; and interruption of air traffic control. Although the private sector will make many of the technical advances in this field, there is much technical work for the Army to do on its own specialized systems.

Conclusion 6-1. Science and technology can and will assist the Army in its homeland security role.

Recommendation 6-1. The Army should focus its funding and research efforts on the high-payoff technologies shown in summary Table 6-1.

TABLE 6-1 High-Payoff Technologies

Function	Technology	Availability[a] (R, N, F)	Multiuse[b] (H, O, C)
Chapter 2	Indications and Warning Technologies		
Perimeter defense and warning	HgCdTe imaging LWIR arrays to fabricate high-performance detector arrays.[c]	R	H, O, C
	Uncooled bolometer arrays utilizing temperature-dependent dielectric constants and operating at room temperature.[c]	R, N	H, O, C
	GaAs quantum well arrays; a type of extrinsic photoconductor in which the bound electrons reside inside the quantum wells instead of on dopant ions.[c]	R, N	H, O, C
	GaN UV detectors for solar blind applications.[d]	F	H, O, C
Biological agent detection	DNA microarrays that can monitor thousands of genes simultaneously.	F	H, O, C
	Combinatorial peptides using massive libraries for screening.	F	H, O, C
	Raman scattering; matches observed Raman spectra against library of predetermined signatures.[e]	N, F	H, O, C
Vapor-phase explosive detectors	Chemical resistors that detect at parts per billion level. Must be close to explosive or chemical, needs improved SNR.[f,g]	N	H, O, C
	Fluorescent polymers that detect at parts per trillion level (in principle). Must be close to explosive or chemical, needs improved SNR. Demonstrated at parts per billion in reliable system.[h]	R, N	H, O, C
	Surface-enhanced Raman spectroscopy that detects at parts per billion. Portable, must be close to explosive.[h]	N, F	H, O, C
	Immunoassay (biosensors) that detects parts per billion. Must be close to explosive. Potential for increased sensitivity.[h]	N, F	H, O, C

Continues

TABLE 6-1 Continued

Function	Technology	Availability[a] (R, N, F)	Multiuse[b] (H, O, C)
Bulk explosive detection	Nuclear quadrupole magnetic resonance (NQR). Low SNR, must be close to explosive, does not require magnets. Produces RF signals characteristic of particular explosives.[g,i]	R, N	H, O, C
	Millimeter-wave radiometry. Potential to provide radiometric images of objects (e.g., explosives) under clothing.[g,j]	N	H, O, C
Cross-cutting detection and tracking	Sensor networking—gathers data from a wide variety of spatially distributed sensors.	N, F	H, O, C
	Sensor fusion—intelligently combines, correlates, and interprets data from distributed sensors.	N, F	H, O, C
	Anomaly detection—examines data from networked sensors to discover patterns, unusual behavior, etc.	N, F	H, O, C
	Surveillance platforms (UAVs, UGVs, UUVs)— small autonomous vehicles for carrying sensor payloads as part of distributed sensor network.	R, F	H, O, C
Cross-cutting perimeter surveillance	IR, RF, acoustic, seismic, etc. techniques that monitor for intrusion into predetermined spaces (encampments, facilities, borders, etc.).	R, N	H, O, C
Cross-cutting capability in miniaturized systems	MEMS—methods for integration of many technologies into microsensors using electronic fabrication technologies.	R, F	H, O, C
	Active-passive sensor suites—suites of lasers and detectors that can query and image as well as perform spectroscopic measurements.	N, F	H, O, C
	Nanofabrication techniques—fabrication of sensing systems at the atomic level.	F	H, O, C

Function	Technology	Availability[a] (R, N, F)	Multiuse[b] (H, O, C)
Chapter 3	Denial and Survivability Technologies		
Perimeter control	X-ray assessment, swimming sensors for rapid detection of LVBs.	N, F	H, O
	Unattended sensor networks, advanced power sources, C2 and secure communication, low-power sensing elements for deployable perimeter control system.	N, F	H, O
	C2 and secure communications, situational awareness tools, area sensors for mobile perimeter system.	F	H, O
Building and facility access control	Smart ID with bioinformation, ID tracking with area authorization, iris ID, liveness tests, auto DNA ID for automatic, high-confidence access control.	F	H, O, C
Structural blast resistance	Prediction of blast and impact loads on and in buildings, bridges, dams, etc.	N, F	H, O, C
	Connection details for steel and concrete structures (new and retrofit construction) to upgrade current approaches for dynamic environments and material behavior.	N	H, O, C
	Methodology to prevent/evaluate potential for progressive collapse.	N (+ university, industry)[k]	H, O, C
	Blast-resistant window concepts, including new glazing-to-frame connections.	N	H, O, C
	Blast-resistant tempered and laminated glass (stiffness, strength enhancement, ductility).	F	H, C
	First-principles analysis techniques to supplement experimental databases for design of windows and structural component retrofits.	N	H, O, C

Continues

TABLE 6-1 Continued

Function	Technology	Availability[a] (R, N, F)	Multiuse[b] (H, O, C)
	Software to include new test and analysis data and techniques for design and retrofit of structures in blast environments.	R, N	H, O, C
	Integration of performance standards with building codes from a multihazard perspective.	N, F	H, O, C
Cybersecurity	IP version 6 to provide ad hoc mobile C&C networks to rapidly reconfigure systems.	N	H, O, C
	Technologies to avoid enemy intrusions, guarantee functionality.	F	H, O
	Technologies to provide alternative C&C after a disaster.	N	H, O
	IP version 6 for networks, universal radio, etc. to allow the Army systems to interoperate with other emergency services.	N	H, O
Chapter 4	Recovery and Consequence Management Technologies		
Command and control	Adaptive integrated multiplexer systems to integrate communications between multiple agencies.	N	H, O, C
	Mobile local broadband networks to pass imagery and communications.	N, F	H, C
	Blue Force Tracking to determine the location of operational personnel and assets from multiple agencies.	N, F	H, O, C
Planning	Decision support aids such as those in the Agile Commander ATD to enhance real-time planning among multiple agencies.	N	H, O
Event assessment	Family of interoperable operational pictures displays that can be shared by operational planners and implementers.	N, F	H, O, C

Function	Technology	Availability[a] (R, N, F)	Multiuse[b] (H, O, C)
	Land mobile robotics that can breach obstacles to implant sensors.	R, N	H, O, C
	Sensor networking and fusion to integrate multiple sensors into a common picture.	N, F	H, O, C
	Real-time damage and contamination modeling to provide attack assessments based on the reports of fused sensor data.	N, F	H, O, C
Force protection	Development of improved protective mask filters and service-life indicators.	R, N	H, O, C
	Development of semipermeable membranes and self-detoxifying material for protective suits.	N	H, O, C
	Vaccine development for protection against biological agents.	N, F	H, O, C
Medical response	Chemical, biological, and radiological triage assessment cards providing C4ISR integration of data, decontamination of the patients and material, tracking of the patients, physical evidence, clothing; chain of custody.	R, N	H, O, C
	C4ISR; on-demand access to expert's network, scenario modeling/procedures to provide remote expert support for the on-site medical personnel; on-demand linkage to medical and scientific information systems, experts, and laboratories.	R, N	H, O, C
	Field-deployable diagnostic, life-support, and emergency surgical systems that can be easily and rapidly deployed; that are resistant to vibration, low environmental quality and electromagnetic interference; and that can be operated efficiently in the presence of chemical, biological, or radiological residuals.	R, N, F	H, O, C

Continues

TABLE 6-1 Continued

Function	Technology	Availability[a] (R, N, F)	Multiuse[b] (H, O, C)
	Field-deployable rapid-assay devices; dynamic meteorologic models of CBRN threats to provide the first responder an assessment of agents and risks for staff and patients; assessment of ongoing environmental risks.	R, N	H, O, C
	Scenario development software based on physiologic and biochemical response to agents.	R, N	H, O
	Hemorrhage, neurological, and respiration stabilizing devices and technologies with a long shelf-life, rapid-acting agents.	R, N	H, O, C
	Vaccines and immunologic factors (including therapeutic applications), counteragents for chemical, biological, and radiological exposure with a long shelf-life, rapid-acting agents.	R, N, F	H, O
	Distributed learning platforms with AI and decision-assisting tools for CBRNE.	R, N, F	H, O
Remediation and decontamination	Development of a process to plan and implement remediation and decontamination for chemical, biological, radiological, and nuclear events.	N	H, C
	Further development and assessment of solutions to clean up chemical and biological contamination.	R, N, F	H, C
Chapter 5	Attribution and Retaliation Technologies		
Detect traffic/ activity abnormality in urban and rural locations	Multisensor fusion.	N	H, O
	Data mining techniques.	N	H, O
	Inference algorithms.	N	H, O
	Redeployable UGS.	F	H, O

Function	Technology	Availability[a] (R, N, F)	Multiuse[b] (H, O, C)
Locate terror cells in areas of heavy foliage	3-D ultrasensitive lidar.	N	O
Defeat covered and concealed targets in rural environment	3-D ultrasensitive lidar.	N	O
	Multisensor fusion techniques.	N	O
Locate gunshots in urban environment	Ultrasensitive acoustics triangulation . system.	F	H, O, C
Enhanced red force (enemy) location in urban environment	Track deconfliction algorithms.	F	O
Situational awareness	Enhanced blue force (friendly) personnel location in urban environment provided by fused GPS, RF, and dead-reckoning hardware and algorithms.	N	H, O, C
Mobility in remote urban environment	Exoskeleton for soldier platform.	F	O, C
	Light, highly survivable, signature-suppressed troop-carrying helicopter.	F	O, C
	Mobile, small-scale robotic breachers for clearing alleys, etc. in urban environment.	N, F	O, C
Remote operations	Reduced usage of signature-producing technologies.	N	H, O
	Advanced composites for lightweight armor protection.	F	H, O, C
	Advanced composites for enhanced vehicle mine protection.	F	H, O, C
	Advanced health and wound monitoring system that integrates blood pressure, heart rate, body temperature, skin penetration sensors.	N, F	H, O, C

Continues

TABLE 6-1 Continued

Function	Technology	Availability[a] (R, N, F)	Multiuse[b] (H, O, C)
Munitions and delivery systems designed for remote urban combat	Nonlethal munitions to include acoustic systems.	N, F	H, O, C
	PSYOP products.	N	O
	UAVs and UGVs designed for urban fire support.	N	H, O, C
Precision insertion and targeting for warheads	Advanced propellants.	N, F	O
	Improved warhead design.	N, F	O

NOTE: AI, artificial intelligence; ATD, Advanced Technology Demonstration; CBRN, chemical, biological, radiological, and nuclear; CBRNE, chemical, biological, radiological, nuclear, and high explosive; C&C, computers and communication; C2, command and control; DARPA, Defense Advanced Research Projects Agency; EO, electro-optical; FOLPEN, foliage penetration; GPS, Global Positioning System; ID, identification; IP, Internet protocol; IR, infrared; lidar, light detection and ranging; LVB, large vehicle bomb; LWIR, long-wave infrared; MEMS, microelectromechanical systems; NSA, National Security Agency; PSYOP, psychological operations; RF, radio frequency; SNR, signal-to-noise ratio; UAV, unmanned air vehicle; UGS, unattended ground sensor; UGV, unmanned ground vehicle; UUV, unmanned underwater vehicle; UV, ultraviolet; 3-D, three-dimensional.

[a]Availability: R, ready (TRL 8-9); N, near-term (TRL 4-7); F, far-term (TRL 1-3).

[b]Multiuse: H, Army homeland security; O, Objective Force; C, civilian (first responders and others).

[c]Westervelt et al. (1991).

[d]DARPA (2002a,b).

[e]NATIBO (2001).

[f]Lewis et al. (1997).

[g]Bruschini and Gros (1997).

[h]Ward et al. (2001).

[i]U.S. Navy (2002).

[j]NRC (1996).

[k]Participation by universities and industry should be sought, because their technology, understanding, experience, and capabilities in this area are advanced, their databases are useful, and they would provide new insight and information to the program and shorten the time frame for development.

REFERENCES

Bruschini, C., and B. Gros. 1997. A Survey of Current Sensor Technology Research for the Detection of Landmines. Available online at <http://diwww.epfl.ch/lami/detec/susdemsurvey.html>. Accessed on September 24, 2002.

DARPA (Defense Advanced Research Projects Agency). 2002a. Semiconductor Ultraviolet Optical Sources (SUVOS) Available online at http://www.darpa.mil/mto/suvos/index.html. Accessed on October 2, 2002.

DARPA. 2002b. Solar Blind Detectors. Available online at http://www.darpa.mil/MTO/SBD/index.html. Accessed on October 2, 2002.

Lewis, N.S., M.C. Lonergan, E.J. Severin, B.J. Doleman, and R.H. Grubbs. 1997. Array-based vapor sensing using chemically sensitive carbon black-polymer resistors. Pp. 660-670 in Detection and Remediation Technologies for Mines and Minelike Targets II, Proceedings of SPIE, vol. 3079, A.C. Dubey and R.L Barnard, eds. Bellingham, Wash.: The International Society for Optical Engineering.

NATIBO (North American Technology and Industrial Base Organization). 2001. Biological Detection System Technologies Technology and Industrial Base Study, February, Available online at <http://www.dtic.mil/natibo/>. Accessed on September 23, 2002.

NRC (National Research Council). 1996. Airline Passenger Security Screening: New Technologies and Implementation Issues. Washington, D.C.: National Academies Press.

U.S. Navy. 2002. Department of the Navy Explosive Detection Equipment-Explosives. Available online at <http://explosivedetection.nfsec.navy.mil/explosives./htm>. Accessed on September 24, 2002.

Ward, K.B., A. Ervin, J.R. Deaschamps, and A.W. Kusterbeck. 2001. Force protection: Explosives detection experts workshop, NRL/MR-MM/6900—01-8564, CDROM. Arlington, Va.: Office of Naval Research.

Westervelt, R., J. Sullivan, and N. Lewis. 1991. Imaging Infra-red Detectors. JASON report number JSR-91-600. McLean, Va.: Mitre Corporation.

7

Complete List of Findings, Conclusions, and Recommendations

Finding 1-1. Homeland security is an important extension of the Army's historical role of providing military support to civilian authorities. The Army will be called on to assist the lead federal agency, the Department of Homeland Security, in meeting a wide range of demands for consequence management and recovery of public order and critical services.

Finding 1-2. The Army National Guard, given its historical mission and flexibility, geographic dispersion, dual-mission capabilities, and frequent association with local agencies, is the key Army asset to meet homeland security demands and can be augmented as necessary with special capabilities from the Army Reserve and the active Army.

Finding 1-3. There are many similarities between military operations involving allied or coalition forces and operations involving civilian emergency responders.

Conclusion 1-1. Many of the technological requirements for homeland security will be important for the Objective Force.

Recommendation 1-1. To optimize current science and technology efforts, the Army should take advantage of potential transferability between technologies for homeland security and those for the Objective Force.

Conclusion 1-2. There needs to be better means to coordinate the homeland security science and technology efforts of the Department of Defense and those of the various civilian agencies.

Recommendation 1-2. The Army should encourage better coordination of the disparate homeland security science and technology efforts.

Conclusion 1-3. Homeland security technologies developed by the Army could be of great benefit to the private sector and to other government agencies.

Recommendation 1-3. The Army should facilitate technology transfer in order to allow the private sector and other government agencies to exploit the homeland security technologies it develops.

Conclusion 1-4. The ability to rapidly deploy a capability-based task force in support of either the homeland security mission or an Objective Force mission will become even more critical.

Recommendation 1-4a. The Army should investigate the technologies necessary to put together on the fly the force packages necessary to meet the requirements of both homeland security and the highly deployable Objective Force.

Recommendation 1-4b. Given the time lag associated with training personnel and leadership to use new technology, now is the time to start dealing with these issues in the context of homeland security, so that they are well honed by the time the Objective Force is fielded.

Conclusion 1-5. The Army National Guard does not appear to play a direct role in defining the critical requirements associated with homeland security.

Recommendation 1-5. The Army National Guard's homeland security role must be considered in the development of the Army Science and Technology Master Plan, and resources for these requirements applied as appropriate in developing the Department of the Army Master Priority List.

Conclusion 1-6. Command, control, communications, computers and intelligence, surveillance, and reconnaissance (C4ISR) is of supreme importance and will apply to a greater or lesser extent in each of the four operational areas in both homeland security and the Objective Force.

Conclusion 2-1. In conducting the survey it was often difficult to obtain authoritative and certified data on the real-world performance of many of the indicators and warning sensors in use or in development. This difficulty also applied to data on sensitivity and noise characteristics.

Recommendation 2-1. It is critically important that all sensors not only be well characterized at the point of purchase but also be regularly rechecked by competent technicians. Software used to integrate disparate sensors should be well documented and checked against standardized problems.

Conclusion 2-2. Technologies should be pursued that (1) deny theft or diversion by maintaining real-time inventory control, then tracking if control is lost or (2) reduce the utility of such equipment to terrorists. Incorporation of detection markers and identification taggants into all legitimately manufactured low-vapor-pressure explosives will assist in both detection and forensic analysis.

Recommendation 2-2. An international convention requiring the incorporation of detection markers and identification taggants should be sought.

Conclusion 2-3. The physical detection of dangerous packaged materials (nuclear weapons, radiological weapons, chemical weapons, biological weapons, and explosive weapons) is an extremely difficult and stressing task, even when the materials are forced through choke points.

Conclusion 2-4. A purely technical solution to the indications and warning problem based upon sensors, even networked sensors, is unlikely. Establishing the proper interrelationships among the sensor networks and the broader intelligence collection activity will be crucial for properly queuing the sensor network.

Recommendation 2-4a. The Army should ensure from the outset that the necessary interrelationships among the sensor networks and the broader intelligence collection activity are established and maintained as a coherent undertaking.

Recommendation 2-4b. Army science and technology should aggressively seek out and invest in those cross-cutting sciences and technologies that will benefit both the Objective Force and the homeland security requirement to detect weapons of mass destruction.

Conclusion 3-1. The current database describing injuries and fatalities due to blast-related terrorist activities is sparse.

Recommendation 3-1. To gather valuable and perishable medical and other forensic data, the Army should support the establishment of rapid response data-gathering teams to investigate bombing attacks that may occur in the future. The data collected by these teams should be integrated with information from past events and made available to researchers and practitioners in emergency medicine, injury epidemiology, search and rescue, architecture, and engineering.

Conclusion 3-2. Heating, ventilation, and air conditioning systems can be improved and integrated with architectural/civil design features for both new buildings and retrofits to provide better resistance to chemical, biological, and radiological attacks.

Recommendation 3-2. The Army should monitor and integrate new heating, ventilation, and air-conditioning technologies developed by the Defense Advance Research Projects Agency and other organizations into building and infrastructure design and retrofit guidelines. These technologies include detection, neutralization, filtration, and active ventilation defenses.

Conclusion 3-3. Research currently being conducted by universities in window/glass behavior and structural response through failure in dynamic environments can help to improve the blast resistance of key structures.

Recommendation 3-3. The Army should continue to survey and evaluate relevant ongoing university research with the objective of identifying and synthesizing technology that could improve the performance of buildings in a blast environment, and it should also consider inviting universities to directly participate in the research effort.

Conclusion 3-4. As the Army becomes more dependent on computer-based systems, cybersecurity becomes more of an issue.

Recommendation 3-4a. The Army should partner with other agencies and the commercial sector to develop and adopt the appropriate tools and protocols for the protection of its own computer and communication systems.

Recommendation 3-4b. The Army should continue to review its cybersecurity procedures to assure that the best practices from the community are adopted on an ongoing basis.

Conclusion 3-5. Even if the attack does not directly inflict physical or cyberdamage on computer and communication systems, the public systems may become overloaded. Since the first responders often use components of public systems, command, control, communications, computers, intelligence, surveillance, and reconnaissance (C4ISR) may be a significant problem in the aftermath.

Recommendation 3-5a. Whether through the Army National Guard or active or reserve Army units, the Army should play a major role in providing emergency command, control, communications, computers, intelligence, surveillance, and reconnaissance (C4ISR) in the event of a major natural or terrorism disaster because it has both the skill set and the equipment to provide such services in hostile environments.

Recommendation 3-5b. Equipment and trained personnel should be available to provide vital information and communications for interoperable command, control, communications, computers, intelligence, surveillance, and reconnaissance (C4ISR) in the case that civilian systems are seriously impaired in an emergency event.

Conclusion 4-1. A new national emergency response command, control, and communications system for homeland security must be developed and fielded to meet the demands of the emerging threats, particularly to integrate the response to chemical, biological, high explosive, radiological, and nuclear weapons. This system must be compatible with developments in the new Department of Homeland Security, the U.S. Northern Command, and state and local entities. Current Army science and technology thrusts and programs that are integral to the Objective Force can be adapted for the new national system.

Recommendation 4-1. To facilitate the development and fielding of an integrated command-and-control system for homeland security, the Army should initiate or continue research that permits the earliest possible fielding of investigate deployable communications packages equipped with universal multiplexer capability to facilitate command and control across the vast, and disparate, array of agencies that will respond to incidents and events.

Conclusion 4-2. Rapid assessment of the effects of natural disasters and attacks using chemical, biological, high explosive, radiological, and nuclear weapons is essential to mitigate the damage, save lives, and restore order. To some degree, the process for event assessment is similar to that used by the Objective Force in building a common operational picture; however, different sensors and analytical processes will be used.

Recommendation 4-2. The Army should conduct research on processes and systems to facilitate the event assessment process. It should support high-priority research such as sensor networking and fusion to merge reports from disparate sensors into a common picture.

Conclusion 4-3. An aggressive, continuing science and technology program across the spectrum of technologies needed for individual and collective protection is necessary for the Army and civilian emergency responders.

Recommendation 4-3. The Army's research and development across the spectrum of technologies needed for individual and collective protection against the effects of weapons of mass destruction for the Army and civilian emergency responders should be continued.

Conclusion 4-4. The new challenges for recovery and consequence management include triage, tracking, and treatment of mass casualties following an event involving weapons of mass destruction. The scale of such an event and the need to conduct an orderly treatment process in the presence of chemical, biological, radiological, or nuclear contamination is daunting. In all likelihood, the nation's military, including the Army, will be called on to play a significant role in this activity.

Recommendation 4-4a. The Army should expand its research in the area of triage, tracking, and treatment of mass casualties.

Recommendation 4-4b. The Army should ensure development of individual triage assessment for mass casualties from events involving weapons of mass destruction.

Recommendation 4-4c. The Army should ensure the development of a process to leverage information technology to effectively conduct mass casualty triage, tracking, and treatment following such an event. The process development should incorporate remote decision support systems that can be integrated with civilian systems, and a tracking system.

Conclusion 4-5. The processes for decontamination following chemical, biological, radiological, nuclear, or even large explosive events need to be expanded. Rapid remediation of the areas involved in such an event will be necessary to limit casualties and to restore critical services. Expanded Army science and technology can contribute significantly to process development and to finding decontamination materials to assist the activity.

Recommendation 4-5a. Army science and technology should concentrate on the further development of a process to plan and implement remediation and decontamination for chemical, biological, radiological, and nuclear events. This process must be capable of being conducted in real time based on limited information.

Recommendation 4-5b. Army science and technology should concentrate on the further development of decontamination solutions for chemical, biological, radiological, nuclear, or even large explosive events.

Conclusion 5-1. Lack of mobility in an urban environment is a critical disadvantage that can result in survivability challenges.

Recommendation 5-1. The Army should continue and enhance current research and development to focus on mobility operations in the urban environment, to include exploration of small, mobile armored carriers for use in urban environments and mini-breachers to clear streets and alleyways.

Conclusion 5-2. Precision and lethality of weapons are critical issues the Army should address to improve fire support for operations in urban environments.

Recommendation 5-2. The Army should modify current systems or develop new systems, along with appropriate munitions, that are specifically designed for extremely precise fire support in urban environments.

Conclusion 5-3. Several capabilities and technologies being developed by the Army would be extremely useful for the civilian first responder, for example the situational awareness Blue Force Tracking and health monitoring system.

Recommendation 5-3. The Army should make technologies such as the situational awareness Blue Force Tracking program and the health monitoring system available to the Department of Homeland Security, which will consider whether or not they can be adapted for civilian use.

Conclusion 5-4: A very sophisticated situational awareness system, with highly accurate Blue Force Tracking in an urban environment, although difficult to construct due to complex radio frequency characteristics and the degree of accuracy required, will provide the soldier and civilian emergency responders a very powerful tool in the war against terrorism.

Recommendation 5-4: The Army should continue to develop a robust soldier situational awareness system begun in Land Warrior that provides a real-time, fused information system.

Conclusion 5-5. Terrorist cell tracking and surveillance in the urban environment and in rugged terrain are extremely difficult as they rely on a very small signal against a large background.

Recommendation 5-5. The Army should adopt a tiered approach to the problem of terrorist cell tracking and surveillance in the urban environment and in rugged terrain, first increasing sensor sensitivity, then networking and fusing sensors, and, finally, fusing information from disparate sources.

Conclusion 6-1. Science and technology can and will assist the Army in its homeland security role.

Recommendation 6-1. The Army should focus its funding and research efforts on the high-payoff technologies shown in summary Table 6-1.

APPENDIXES

Appendix A

Biographical Sketches of Committee Members

John W. Lyons, NAE, *Chair*, consultant and retired director of the Army Research Laboratory (ARL), is a Ph.D. physical chemist. He served in research and development positions with the Monsanto Company for 18 years. In 1973 he joined the Commerce Department's National Bureau of Standards (NBS). At NBS, Lyons was the first director of the Center for Fire Research. In 1990 Dr. Lyons was appointed by President George H.W. Bush to be the ninth director of NBS, by that time renamed the National Institute of Standards and Technology (NIST). In September 1993, he was appointed the first permanent director of ARL. At ARL, Dr. Lyons managed a broad array of science and technology programs. He has served on many boards and commissions, inter alia, the Federal Advisory Commission on Consolidation and Conversion of Defense Research and Development Laboratories. He currently serves on two boards of visitors at the University of Maryland. He is a member of the National Research Council's Board on Army Science and Technology, as well as a member of a congressionally chartered committee at the National Defense University to study the potential effectiveness of the DoD laboratories in the transformed military of the future. Dr. Lyons was elected to the National Academy of Engineering in 1985. He is a fellow of the American Association for the Advancement of Science and of the Washington Academy of Science and is a member of the American Chemical Society and of Sigma Xi.

George Bugliarello, NAE, is presently chancellor of Polytechnic University, Brooklyn, New York. Dr. Bugliarello, a former president (1973-1994) of Polytechnic, an engineer and educator whose background ranges from biomedical engineering to fluid mechanics, computer languages, socio-technology, and science policy, is a leader of the Urban Security Initiative at Polytechnic. A member of the National Academy of Engineering (NAE) and the Council on Foreign Relations and a founding fellow of the American Institute of Medical and Biological Engineering, he is a past president of the Sigma Xi, the scientific research society, and holds honorary lifetime membership in the National Association for Science, Technology, and Society (NASTS). He has served as both member and chair of several National Research Council committees, among the latest of which were chairmanship of the Committee on Alternative Technologies to Replace Anti-Personnel Landmines, and membership in the Committee on Human Rights of the National Academy of Sciences, NAE, and the Institute of Medicine (IOM). He is currently a member of the National Research Council Committee on Counterterrorism Challenges for Russia and the United States. Dr. Bugliarello's international experience includes consultancies abroad for United Nations Economic and Social Commission and the Organization for Economic Cooperation and Development, being the U.S. member of the Science for Peace Steering Group of the North Atlantic Treaty Organization (NATO), and, previously, of NATO's Science for Stability Steering Group.

Timothy Coffey currently holds the Edison Chair at the Center for Technology and National Security Policy at the National Defense University and is a senior research scientist at the University of Maryland. He graduated from the Massachusetts Institute of Technology in 1962 with a B.S. degree in electrical engineering and obtained his M.S. (1963) and Ph.D. (1967), both in physics, from the University of Michigan. During his graduate career, Dr. Coffey worked as a research assistant at the University of California (1963-1964), a research physicist at the Air Force Cambridge Research Laboratories (1964-1965), and a teaching fellow and research assistant in physics at the University of Michigan (1965-1966). As a scientific consultant for EG&G, Inc. (1966-1971), he was involved in investigations in theoretical and mathematical physics. Dr. Coffey joined the Naval Research Laboratory in 1971 as head of the Plasma Dynamics Branch, Plasma Physics Division. In this position, he directed research in the simulation of plasma instabilities, the development of multidimensional fluid and magnetohydrodynamic codes, and the development of computer codes for treating chemically reactive flows. In 1975, he was named superintendent, Plasma Physics Division; he was appointed associate director of research for General Science and Technology on January 1, 1980. On November 28, 1982, he was named Director of Research. In October 2001 Dr. Coffey retired from the Naval Research Laboratory and joined the University of Maryland. Dr. Coffey conducted research on the theory of nonlinear oscillations and played a major role in the national program on high-altitude nuclear effects. The author or coauthor of over 70 publications and reports, he has made several fundamental contributions to the theory of electron beam/plasma interaction and to the understanding of plasma processes in

Earth's ionosphere. Dr. Coffey is a fellow of the American Physical Society, of the Franklin Institute, and of the Washington Academy of Science and a member of the American Institute of Physics, of the American Association for the Advancement of Science, and of Sigma Xi. In 1981, he was awarded the Presidential Rank of Meritorious Executive. He was awarded the Presidential Rank of Distinguished Executive in 1987 and 1994. In 1991, Dr. Coffey was the recipient of the Delmer S. Fahrney Medal and received the Department of Defense Distinguished Civilian Service Award. On March 14, 1996, he was awarded the Senior Executives Association Professional Development League's 1995 Executive Excellence Award for Distinguished Executive Service. In August 2000, he was awarded the Navy's prestigious Captain Robert Dexter Conrad Award. Dr. Coffey was selected by Irish American Magazine as one of the top 100 Irish Americans for the year 2000. Upon his retirement from the Naval Research Laboratory, he was awarded the Laboratory's Lifetime Achievement Award.

Stephen W. Drew, NAE, currently spreads his efforts between professorships at Princeton and Cambridge Universities and consultancies with a variety of pharmaceutical and biotechnology organizations. Until 2000, he worked with Merck & Company, Inc., in a series of increasingly responsible positions culminating as the distinguished senior scientist. Dr. Drew received his Ph.D. in biochemical engineering from the Massachusetts Institute of Technology. A member of the NAE, he has served in several capacities within the NAE itself and assisted numerous National Research Council committees.

Mitra Dutta currently serves as professor and head of electrical and computer engineering, as well as adjunct professor of physics, at the University of Illinois at Chicago. She received her B.Sc. and M.Sc. degrees from India (Delhi University) and M.S. and Ph.D. degrees from the University of Cincinnati, in Ohio. She has held appointments at the College of Arts and Sciences at Kingston, Jamaica, in the West Indies, postdoctoral appointments at Purdue University and the City College of New York, and adjunct professor appointments at Rutgers University, the University of Maryland, North Carolina State University, the University of North Carolina at Chapel Hill, and Brookhaven National Laboratory. She worked for 10 years at the U.S. Army Research Laboratory in various capacities, and prior to joining the faculty of the University of Illinois, Dr. Dutta served in a senior executive service position in the Army Research Office (ARO), now a component of the U.S. Army Research Laboratory. She has authored or coauthored over 350 publications and presentations, holds 26 U.S. and Canadian patents, has coedited two books and is a coauthor of a third. She is a fellow of the Institute of Electrical and Electronics Engineers (IEEE) and of the Optical Society of America and was the recipient of the IEEE Harry Diamond Award in 2000. Her interests include the electrical, optical, and mechanical properties of nanostructures, quantum transport, solid-state electronics and optoelectronics, phonons in nanostructures, theory of nanodevices, and applications of nanoscale structures and devices in electrical engineering and bioengineering.

Frederick L. Frostic is currently a principal with Booz Allen Hamilton. Prior to joining Booz Allen, he served as Deputy Assistant Secretary of Defense for Requirements and Plans, where he was responsible for preparing the Defense Planning Guidance, supervised the Defense Department's response to the congressionally mandated Commission on Roles and Missions, and conducted crisis planning, plans reviews, and force structure analysis. Recently, he was the project manager of a group providing research to the U.S. Commission on National Security/21st Century (Hart-Rudman Commission). In this effort, his team wrote the implementation plan for the commission's recommendations on homeland security. Additionally he was the project manager to provide research support to the Presidential Commission on Critical Infrastructure Protection. Mr. Frostic, a graduate of the Air Force Academy, earned an M.S. in engineering from the University of Michigan in 1971 and conducted postgraduate work in aerospace engineering until 1976.

C. William Gear, NAE, is president emeritus of the NEC Research Institute. Prior to joining NEC, he was head of the Department of Computer Science and professor of computer science and applied mathematics at the University of Illinois at Urbana-Champaign. His research expertise is in numerical analysis and computational software. Dr. Gear is a member of the National Academy of Engineering and a fellow of the American Academy of Arts and Sciences, IEEE, the American Association for the Advancement of Science, and the Association for Computing Machinery. He served as president of the Society for Industrial and Applied Mathematics and was the recipient of the ACM SIGNUM George E. Forsythe Memorial Award and Fulbright and Johnson Foundation Fellowships.

Arthur H. Heuer, NAE, is University Professor and the Kyocera Professor of Ceramics at Case Western Reserve University. His interests include microelectromechanical systems (MEMS), phase transformations and dislocations in ceramics, rapid prototyping, structure/property/function studies of biological ceramics (teeth, shell, and bones), and the applications of biological processes to the processing of advanced ceramics. He received a B.S. in chemistry from the City College of New York, a Ph.D. in applied science, and a D.Sc. in physical ceramics from the University of Leeds, England. He is a member of the National Academy of Engineering and an external member of the Max Planck Institute for Material Science, Stuttgart, Germany.

Howard S. Levine is a principal with Weidlinger Associates, Inc. His responsibilities include analysis of ground motion and structural response from nuclear and conventional explosions, aircraft impact, and earthquakes. Dr. Levine is currently leading development and analysis efforts in air blast, fragment, and ground shock loading of hardened reinforced concrete structures, deep tunnels in rock, and aboveground industrial structures subjected to conventional weapons effects. He received a B.S. in aerospace engineering, an M.S. in applied mechanics, and a Ph.D. in applied mechanics, all from the Polytechnic Institute of Brooklyn. Dr. Levine has numerous affiliations that include the American Society of Mechanical Engineers, the American Society of Civil Engineers, Tau Beta Pi, Sigma Xi, and Sigma Gamma Tau.

Joseph P. Mackin, a retired Army Acquisition Corps colonel, is currently president of E-OIR Technologies, Inc., a high-technology sensor applications company in Virginia. He has an extensive background in sensors, having served in many DoD sensor development and acquisition assignments such as deputy division director of the Laser Division at the Night Vision and Electronic Sensors Directorate; as product manager for the Army's second generation FLIR (thermal imager) for the Abrams Tank and the Bradley Fighting Vehicle; and as the Director of Special Programs on the staff of the Army Acquisition Executive. Since retiring from the Army and prior to accepting his current position, he worked at MIT/Lincoln Laboratory as an assistant group leader in the Sensors Applications Group, where he was the technical lead for the Deputy Undersecretary of Defense for Science and Technology's (DUSD S&T) Smart Sensor Web program. His education includes a B.S. from the U.S. Military Academy at West Point, an M.S. in physics (electro-optics) from the Naval Postgraduate School, and a Ph.D. in physics (atomic and lasers) from the Massachusetts Institute of Technology. He is also a graduate of the Defense Systems Management College.

Jack N. Merritt serves concurrently as chairman of the Marshall Legacy Institute, director and vice chairman of the Atlantic Council of the United States, director and vice chairman of the George C. Marshall Foundation, and is on the Board of Visitors for the International Center of the University of Oklahoma. A retired U.S. Army general, he was most recently the president and chief operating officer of the Association of the United States Army. General Merritt has had a long and distinguished military career, during which he progressed from the grade of private to four-star general. A former Director of the Joint Staff and Commandant of the United States Army War College, General Merritt's final assignment prior to military retirement was as the U.S. military representative to NATO. He received a B.M.S. from the University of Nebraska at Omaha, an M.S. in business administration from the George Washington University, and was a graduate of the Industrial College of the Armed Forces.

Thomas E. Mitchell is vice president of Gray Hawk Systems, Inc., in Alexandria, Virginia, where he leads the Operations, Intelligence, and Security Division.

Mr. Mitchell, a retired U.S. Army colonel, is a business executive with an extensive background in special operations, crisis response, consequence management, force protection, and critical infrastructure protection. He serves in a strategic consultative role as a member of the Business Advisory Council of the Lexington Institute in Arlington, Virginia, and the Gray Hawk corporate board of directors. Mr. Mitchell received a B.S. from the University of Delaware and an M.P.A. from Jacksonville State University in Alabama. Additionally, he is a graduate of the Army's Advanced Operational Studies War College Fellowship Program, School of Advanced Military Studies.

K. David Nokes currently serves as vice president of the National Security and Arms Control Division at Sandia National Laboratories. He has extensive experience in the design of nuclear weapon systems, arms control, intelligence, and other national security activities. He served as the Special Scientific Advisor to the Assistant to the Secretary of Defense (Atomic Energy), providing advice on nuclear weapon safety, security, and reliability issues. After the break-up of the Soviet Union, he initiated dialogue with and developed programs of cooperation with the nuclear weapon design laboratories of the former Soviet Union, including programs to safeguard their nuclear materials and weapons. In the aftermath of the September 11, 2001, attacks, Mr. Nokes was designated as the Sandia point of contact for Sandia's role in internal and external strategies for engaging Sandia's technology base in problems associated with homeland security and combating terrorism. Mr. Nokes has an M.S. in applied mechanics and an M.S. in computer science and electrical engineering.

Dennis J. Reimer is director of the National Memorial Institute for the Prevention of Terrorism, Oklahoma City. The Institute is dedicated to preventing, reducing, and mitigating the effects of terrorism, with particular emphasis on the role of first responders. A retired U.S. Army general, he was most recently the 33rd Chief of Staff of the Army. He holds a B.S. from the U.S. Military Academy at West Point and a master's degree from Shippensburg State College.

Eugene Sevin, NAE, and a National Associate of the Academies, is a consultant on nuclear and conventional weapons effects, hardened facility design, and computational structural mechanics. He works with the Office of the Secretary of Defense and the Defense Threat Reduction Agency (DTRA) on matters related to target vulnerability, blast mitigation, and high-performance computing in structural mechanics. Dr. Sevin was responsible for experimental research at the Defense Nuclear Agency (now DTRA) and established DTRA's high-performance computing center at Los Alamos National Laboratory. He served as director of space and missiles in the Office of the Undersecretary of Defense (Acquisition). Dr. Sevin received a B.S. in mechanical engineering from the Illinois Institute of Technology, an M.S. in mechanical engineering from the California

Institute of Technology, and a Ph.D. in applied mechanics from the Illinois Institute of Technology.

Annette L. Sobel is a distinguished member of the technical staff of Sandia National Laboratories, New Mexico, and Chemical and Biological Warfare analyst. She has 13 years of advanced technology development and unconventional threat analysis expertise focused on applications of biotechnology and information technologies in support of chemical-biological countermeasures and in the field of human factors/systems engineering (e.g., critical decision making under stress) domains. She is a Brigadier General in the U.S. Air Force Reserve and the Special Assistant for Weapons of Mass Destruction and Civil Support to the Chief of the National Guard Bureau. Her work has emphasized information analysis, advanced systems for mission rehearsal and training, human performance enhancements, and technology transition to field operational environments. She has 11 years of military command experience, including combat and chemical-biological warfare medical response unit commands. Dr. Sobel earned an M.D. at Case Western Reserve University, with specialization in family medicine at Duke University Medical Center. She has an M.S. in aerospace medicine with an emphasis on human factors engineering from Wright State University and a B.S. with high honors. She was a Founder's Scholar in Chemistry and Computer Science at Cook College, Rutgers University. She is a member of the Defense Intelligence Agency's advisory board.

Michael F. Spigelmire is a consultant on crisis response, consequence management, and force protection. A retired U.S. Army lieutenant general, he has had a military career with a unique blend of conventional and special operations assignments. General Spigelmire commanded the U.S. Army's Special Operations Command and then the VII Corps in Germany. Upon retirement, he was deputy director of operations for the Atlanta Committee for the Olympic Games. This brought him into close contact with municipal, state, and federal officials. General Spigelmire holds a B.S. in political science from Loyola College and an M.A. in international relations from Georgetown University. Additionally, he has completed the U.S. Army Command and Staff College and the U.S. Army War College. General Spigelmire is currently the senior mentor for the Crisis Response, Consequence Management Senior Seminar, sponsored by the Joint Special Operations University and the Air Force Special Operations School, Hurlburt Field, Florida.

Appendix B

Committee Meetings

FIRST MEETING

May 14-16, 2002
Warrenton, Virginia

Meeting objectives: National Research Council introduction, complete administrative actions, including committee introductions and composition/balance/bias discussions for members of committee and report procedures, discuss statement of task with sponsor, discuss draft report outline, discuss project plan and report realization, discuss scenarios, review illustrative technologies, make writing assignments, and confirm objectives, location, and dates for the next two committee meetings.

Presenters

Potential Scenarios
Dennis VanDerlaske, ASAALT

Sponsor Discussion Time
John Parmentola, Director of Research and Laboratory Management

DoD's Consequence Management Role in Homeland Security
Kathy Condon, Special Assistant to the Secretary of the Army for Military Support, Office of the Secretary of the Army

Army's Role in Homeland Defense
Gregory J. Bozek, Army War Plans, DAMO-SSW, Headquarters, Department of the Army

Indications and Warning Technologies
Robert Foresta, Branch Chief, U.S. Army CECOM, 12WD
Richard Smarjewski, U.S. Army SBCCOM

Indications and Warning Technologies
Fenner Milton, Director, U.S. Army CECOM ARDEC Night Vision and Electronic Sensors Directorates

Survivability and Denial Technologies
Chuck Kimsey, Kay Blankenship, Richard Smarjewski, U.S. Army SBCCOM
Reed Mosher, U.S. Army Corps of Engineers, Engineer Research and Development Center

Attribution and Retaliation Technologies
Raymond Filler, U.S. Army CECOM C2D
Larry Bovino, U.S. Army CECOM 12WD, Radar Systems Branch
Edward Kierman, Project Leader, U.S. Army CECOM
LTC Kathy DeBolt, Commander, U.S. Army Intelligence Center

Consequence Management and Recovery Technologies
Richard Smarjewski, U.S. Army SBCCOM
Bob Welch, U.S. Army Engineer Research & Development Center
Robert Foresta, Branch Chief, U.S. Army CECOM, 12WD

National Academies' Efforts Concerning Terrorism
Douglas C. Bauer, Director, Counterterrorism Coordination, National Research Council

Consequence Management and Recovery Technologies
LTC Harold Modrow, USAMMDA
Andrzej Miziolek, Propulsion Science Branch, U.S. Army Research Lab
Richard Smarjewski, U.S. Army SBCCOM

SECOND MEETING

June 24-26, 2002
Washington, D.C.

Meeting objectives: Complete composition/balance/bias discussions for committee members, preview additional illustrative technologies, discuss scenarios, discuss project plan and report realization, discuss concept draft, make additional writing assignments, confirm objectives, location, and dates for the next two committee meetings.

Presenters

RAND Scenarios
Randy Steeb, RAND

Posse Comitatus and Other Legal Issues
Joseph R. Barnes, Brig Gen, U.S. Army (retired), Former Assistant Judge Advocate General for Civil Law and Litigation

Preliminary Army Doctrine for Homeland Defense
Larry Heystek, USA Training and Doctrine Command

Role of the Army National Guard in Homeland Defense
Colonel Jeff W. Mathis, III, National Guard Bureau

Technology Briefing—Technology in Support of Recovered Chemical Warfare Materiel
David Hoffman, Office of the Program Manager for the Demilitarization of Chemical Weapons

THIRD MEETING

July 24-25, 2002
Washington, D.C.

Meeting objectives: Complete composition/balance/bias discussions for committee members, preview additional illustrative technologies, discuss project plan and report realization, discuss first full message draft, make additional writing assignments, confirm objectives, location, and dates for the next committee meeting.

Presenters

Cybersecurity
Herbert S. Lin, Senior Scientist, Computer Science and Telecommunications Board

FOURTH MEETING

August 27-28, 2002
Washington, D.C.

Meeting objectives: Discuss project plan and report realization, discuss concurrence draft, and discuss review process.

Presenters

None

Appendix C

Criteria for Technology Readiness Levels

TABLE C1 Criteria for Technology Readiness Levels[a]

TRL	Task Accomplished	Description
1	Basic principals observed and reported	Lowest level of technology readiness. Scientific research begins to be translated into applied research and development. Examples might include paper studies of a technology's basic properties.
2	Technology concept or application formulated	Invention begins. Once basic principles are observed, practical applications can be invented. The application is speculative and there is no proof or detailed analysis to support the assumption. Examples are still limited to paper studies.
3	Analytical and experimental critical function or characteristics proof of concept	Active research and development are initiated. These include analytical studies and laboratory studies to physically validate analytical predictions of separate elements of the technology. Examples include components that are not yet integrated or representative.
4	Component or breadboard validation in laboratory environment	Basic technology components are integrated to establish that the pieces will work together. This is relatively "low fidelity" compared with the eventual system. Examples include integration of ad hoc hardware in a laboratory.

Continues

[a]Adapted from Army Science and Technology Master Plan.

TABLE C1 Continued

TRL	Task Accomplished	Description
5	Component or breadboard validation in relevant environment	Fidelity of breadboard technology increases significantly. The basic technological components are integrated with reasonably realistic supporting elements so that the technology can be tested in a simulated environment. Examples include high-fidelity laboratory integration of components.
6	System/subsystem model or prototype demonstration in a relevant environment	Representative model or prototype system, which is well beyond the breadboard tested for TRL 5, is tested in a relevant environment. Represents a major step up in a technology's demonstrated readiness. Examples include testing a prototype in a high-fidelity laboratory environment or in a simulated operational environment.
7	System prototype demonstration in an operational environment	Prototype near or at planned operational system. Represents a major step up from TRL 6, requiring the demonstration of an actual system prototype in an operational environment, such as in an aircraft, vehicle, or space. Examples include testing the prototype in a testbed aircraft.
8	Actual system completed and flight qualified through test and demonstration	Technology has been proven to work in its final form and under expected conditions. In almost all cases, this TR represents the end of true system development. Examples include developmental test and evaluation of the system in its intended weapon system to determine if it meets design specifications.
9	Actual system flight proven through successful mission operations	Actual application of the technology in its final form and under mission conditions, such as those encountered in operational test and evaluation. In almost all cases, this is the end of the last bug-fixing aspects of true system development. Examples include using the system under operational mission conditions.

Appendix D

Federal Response Plan Responsibilities

The Federal Response Plan[1] (FRP) outlines how the federal government implements the Robert T. Stafford Disaster Relief and Emergency Assistance Act, as amended, to assist state and local governments when a major disaster or emergency overwhelms their ability to respond. What follows is an extract from the Terrorism Incident Annex of the FRP, describing the responsibilities of various federal agencies.

V. RESPONSIBILITIES

A. Department of Justice

Presidential Decision Directive (PDD) 39 validates and reaffirms existing lead agency responsibilities for all facets of the U.S. counterterrorism effort. The Department of Justice is designated as the overall lead federal agency (LFA) for threats of acts of terrorism that take place within the United States until the Attorney General transfers the overall LFA role to the Federal Emergency Management Agency (FEMA). The Department of Justice delegates this overall

[1]Federal Response Plan (Federal Emergency Management Agency. 1999. Federal Response Plan, 9230.1-PL, April. Available online at <http://www.fema.gov/rrr/frp/>. Accessed on December 3, 2002) and its Terrorism Incident Annex (Federal Emergency Management Agency. 1999. Terrorism Incident Annex, April. Available online at <http://www.fema.gov/rrr/frp/frpterr.shtm>. Accessed on December 3, 2002.)

LFA role to the Federal Bureau of Investigation (FBI) for the operational response. On behalf of the Department of Justice, the FBI will:

1. Consult with and advise the White House, through the Attorney General, on policy matters concerning the overall response;
2. Designate and establish a joint operations center (JOC) in the field;
3. Appoint an FBI on-site coordinator (OSC) to manage and coordinate the Federal operational response (crisis management and consequence management). As necessary, the FBI OSC will convene and chair meetings of operational decision makers representing lead State and local crisis management agencies, federal emergency workers, and lead State and local consequence management agencies in order to provide an initial assessment of the situation, develop an action plan, monitor and update operational priorities, and ensure that the overall response (crisis management and consequence management) is consistent with U.S. law and achieves the policy objectives outlined in PDD-39. The FBI and FEMA may involve supporting Federal agencies as necessary; and
4. Issue and track the status of actions assigned by the overall LFA.

B. Federal Bureau of Investigation

Under PDD-39, the FBI supports the overall LFA by operating as the lead agency for crisis management. The FBI will:

1. Determine when a threat of an act of terrorism warrants consultation with the White House, through the Attorney General;
2. Advise the White House, through the Attorney General, when the FBI requires assistance for a Federal crisis management response, in accordance with the PDD-39 Domestic Deployment Guidelines;
3. Work with FEMA to establish and operate a Joint Information Center (JIC) in the field as the focal point for information to the public and the media concerning the Federal response to the emergency;
4. Establish the primary Federal operations centers for the crisis management response in the field and Washington, DC;
5. Appoint an FBI OSC (or subordinate official) to manage and coordinate the crisis management response. Within this role, the FBI OSC will convene meetings with operational decision makers representing Federal, State, and local law enforcement and technical support agencies, as appropriate, to formulate incident action plans, define priorities, review status, resolve conflicts, identify issues that require decisions from higher authorities, and evaluate the need for additional resources;

6. Issue and track the status of crisis management actions assigned by the FBI; and
7. Designate appropriate liaison and advisory personnel to support FEMA.

C. Federal Emergency Management Agency

Under PDD-39, FEMA supports the overall LFA by operating as the lead agency for consequence management until the overall LFA role is transferred to FEMA. FEMA will:

1. Determine when consequences are "imminent" for the purposes of the Stafford Act;
2. Consult with the Governor's office and the White House to determine if a Federal consequence management response is required and if FEMA is directed to use Stafford Act authorities. This process will involve appropriate notification and coordination with the FBI, as the overall LFA;
3. Work with the FBI to establish and operate a JIC in the field as the focal point for information to the public and the media concerning the Federal response to the emergency;
4. Establish the primary Federal operations centers for consequence management in the field and Washington, DC;
5. Appoint a regional operations center (ROC) Director or federal coordinating officer (FCO) to manage and coordinate the Federal consequence management response in support of State and local governments. In coordination with the FBI, the ROC Director or FCO will convene meetings with decision makers of Federal, State, and local emergency management and technical support agencies, as appropriate, to formulate incident action plans, define priorities, review status, resolve conflicts, identify issues that require decisions from higher authorities, and evaluate the need for additional resources;
6. Issue and track the status of consequence management actions assigned by FEMA; and
7. Designate appropriate liaison and advisory personnel to support the FBI.

D. Federal Agencies Supporting Technical Operations

1. Department of Defense

As directed in PDD-39, the Department of Defense (DOD) will activate technical operations capabilities to support the Federal response to threats or acts of weapon of mass destruction (WMD) terrorism. DOD will coordinate military operations within the United States with the appropriate civilian lead agency (ies) for technical operations.

2. *Department of Energy*

As directed in PDD-39, the Department of Energy (DOE) will activate technical operations capabilities to support the Federal response to threats or acts of WMD terrorism. In addition, the FBI has concluded formal agreements with potential LFAs of the Federal Radiological Emergency Response Plan (FRERP) that provide for interface, coordination, and technical assistance in support of the FBI's mission. If the FRERP is implemented concurrently with the FRP:

a. The Federal On-Scene Commander under the FRERP will coordinate the FRERP response with the FEMA official (either the ROC Director or the FCO), who is responsible under PDD-39 for coordination of all Federal support to State and local governments.

b. The FRERP response may include on-site management, radiological monitoring and assessment, development of Federal protective action recommendations, and provision of information on the radiological response to the public, the White House, Members of Congress, and foreign governments. The LFA of the FRERP will serve as the primary Federal source of information regarding on-site radiological conditions and off-site radiological effects.

c. The LFA of the FRERP will issue taskings that draw upon funding from the responding FRERP agencies.

3. *Department of Health and Human Services*

As directed in PDD-39, the Department of Health and Human Services (HHS) will activate technical operations capabilities to support the Federal response to threats or acts of WMD terrorism. HHS may coordinate with individual agencies identified in the HHS Health and Medical Services Support Plan for the Federal Response to Acts of Chemical/Biological (C/B) Terrorism, to use the structure, relationships, and capabilities described in the HHS plan to support response operations. If the HHS plan is implemented:

a. The HHS on-scene representative will coordinate, through the Emergency Support Function (ESF) #8 — Health and Medical Services Leader, the HHS plan response with the FEMA official (either the ROC Director or the FCO), who is responsible under PDD-39 for on-scene coordination of all Federal support to State and local governments.

b. The HHS plan response may include threat assessment, consultation, agent identification, epidemiological investigation, hazard detection and reduction, decontamination, public health support, medical support, and pharmaceutical support operations.

c. HHS will issue taskings that draw upon funding from the responding HHS plan agencies.

4. Environmental Protection Agency

As directed in PDD-39, the Environmental Protection Agency (EPA) will activate technical operations capabilities to support the Federal response to acts of WMD terrorism. EPA may coordinate with individual agencies identified in the National Oil and Hazardous Substances Pollution Contingency Plan (NCP) to use the structure, relationships, and capabilities of the National Response System as described in the NCP to support response operations. If the NCP is implemented:

a. The Hazardous Materials On-Scene Coordinator under the NCP will coordinate, through the ESF # 10—Hazardous Materials Chair, the NCP response with the FEMA official (either the ROC Director or the FCO), who is responsible under PDD-39 for on-scene coordination of all Federal support to State and local governments.

b. The NCP response may include threat assessment, consultation, agent identification, hazard detection and reduction, environmental monitoring, decontamination, and long-term site restoration (environmental cleanup) operations.